AF363301

Predicting the Future

Predicting the Future—Big Data and Machine Learning

Editor

Fernando Sánchez Lasheras

MDPI • Basel • Beijing • Wuhan • Barcelona • Belgrade • Manchester • Tokyo • Cluj • Tianjin

Editor
Fernando Sánchez Lasheras
Oviedo University
Spain

Editorial Office
MDPI
St. Alban-Anlage 66
4052 Basel, Switzerland

This is a reprint of articles from the Special Issue published online in the open access journal *Energies* (ISSN 1996-1073) (available at: https://www.mdpi.com/journal/energies/special_issues/Big_Data_Machine_Learning).

For citation purposes, cite each article independently as indicated on the article page online and as indicated below:

LastName, A.A.; LastName, B.B.; LastName, C.C. Article Title. *Journal Name* **Year**, *Article Number*, Page Range.

ISBN 978-3-03936-619-4 (Hbk)
ISBN 978-3-03936-620-0 (PDF)

Contents

About the Editor

Fernando Sánchez Lasheras (Ph.D.) received M.Sc. and Ph.D. degrees in industrial engineering from the University of Oviedo, Oviedo, in 2000 and 2008, respectively. After a career of almost 20 years in industry and academia, in 2017, he joined the Department of Mathematics of Oviedo University. His current research interests include applied mathematics and machine learning, with more than 80 papers about these topics.

Preface to "Predicting the Future—Big Data and Machine Learning"

This Special Issue of *Energies* "Predicting the Future—Big Data and Machine Learning" deals with interesting and new topics in the field of energy related to recent advances in machine and deep learning. Two of the papers, "Crude Oil Prices Forecasting: An Approach of Using CEEMDAN-Based Multi-Layer Gated Recurrent Unit Networks" and "Optimizing Predictor Variables in Artificial Neural Networks When Forecasting Raw Material Prices for Energy Production" deal with the topic of price forecasting. Another two, "Short-Term Load Forecasting for CCHP Systems Considering the Correlation between Heating, Gas and Electrical Loads Based on Deep Learning" and "Non-Intrusive Load Monitoring (NILM) for Energy Disaggregation Using Soft Computing Techniques" explore demand management and forecasting. Machine learning methodologies have also proven useful in the management of energy systems. This Special Issue has an article related to this topic, "Energy Multiphase Model for Biocoal Conversion Systems by Means of a Nodal Network". There are another two articles that deal with the interesting topics of the use of energy in industrial applications "Multivariate Analysis to Relate CTOD Values with Material Properties in Steel Welded Joints for the Offshore Wind Power Industry" and the health and safety problems of the workers in this field "Prediction of Health-Related Leave Days among Workers in the Energy Sector by Means of Genetic Algorithms". Finally, the Special Issue contains another study in which the impact of energy applications on the environment is considered "Understanding and Modeling Climate Impacts on Photosynthetic Dynamics with FLUXNET Data and Neural Networks".

Fernando Sánchez Lasheras
Editor

Article

Short-Term Load Forecasting for CCHP Systems Considering the Correlation between Heating, Gas and Electrical Loads Based on Deep Learning

Ruijin Zhu, Weilin Guo * and Xuejiao Gong

Electric Engineering College, Tibet Agriculture and Animal Husbandry University, Nyingchi 860000, China
* Correspondence: gwl@my.swjtu.edu.cn; Tel.: +86-131-9347-4789

Received: 4 August 2019; Accepted: 27 August 2019; Published: 28 August 2019

Abstract: Combined cooling, heating, and power (CCHP) systems is a distributed energy system that uses the power station or heat engine to generate electricity and useful heat simultaneously. Due to its wide range of advantages including efficiency, ecological, and financial, the CCHP will be the main direction of the integrated system. The accurate prediction of heating, gas, and electrical loads plays an essential role in energy management in CCHP systems. This paper combined long short-term memory (LSTM) network and convolutional neural network (CNN) to design a novel hybrid neural network for short-term loads forecasting considering their correlation. Pearson correlation coefficient will be utilized to measure the temporal correlation between current load and historical loads, and analyze the coupling between heating, gas and electrical loads. The dropout technique is proposed to solve the over-fitting of the network due to the lack of data diversity and network parameter redundancy. The case study shows that considering the coupling between heating, gas and electrical loads can effectively improve the forecasting accuracy, the performance of the proposed approach is better than that of the traditional methods.

Keywords: short-term loads forecasting; CCHP systems; convolutional neural network; short-term memory network; dropout layer

---

1. Introduction

With the rapid development of industry, the consumption of energy and other natural resources has increased substantially. How to rationally utilize energy resources and improve the efficiency of energy utilization has become a common concern of all countries in the world. The combined cooling heating, and power system is one of the distributed energy systems, which uses a power station or heat engine to generate useful heat and electricity at the same time. It is arranged near the users on a small scale, decentralized, and targeted manner, and delivers heating energy and electric energy to nearby users according to the users' different needs [1,2]. Compared with conventional centralized power systems, the combined cooling, heating, and power (CCHP) system has lower energy costs, higher energy efficiency, and higher energy availability. Therefore, the CCHP system will become the main form of the integrated energy system [3].

The traditional power system, heating system, and natural gas system are independent of each other, which greatly limits the operating efficiency of these three energy systems. The CCHP system uses gas as an energy source and recycles hot water and high-temperature exhaust gas to improve the comprehensive utilization efficiency of energy [4]. In this case, the power system, heating system, and natural gas system will have a strong correlation, which requires the intelligent control of these three systems at the same time. Accurate prediction of heating, gas, and electrical loads is the basic premise of energy management in CCHP systems and has important theoretical and practical value.

Conventionally, heating, gas and electrical loads forecasting are conducted separately, and this is not suitable for CCHP system where the heating, gas, and electrical loads have strong correlations. Therefore, it is necessary to propose a novel load forecasting approach for the CCHP system that accounts for the correlation of these three loads.

Recently, as an important branch of the field of artificial intelligence, the deep learning technology, has been applied to all popular artificial intelligence areas, including speech recognition, image recognition, big data analysis, etc. [5–7]. Especially, the convolutional neural network (CNN), which is well-known for its strong ability to extract features, has gained enormous attention in the field of image classification and image recognition. The CNN with global spatial information was designed to divide white matter hyperintensities in [8]. To realize image classification, the CNN with five convolutional layers and three fully connected layers was designed to improve the accuracy in [9]. The phase-functioned network, a maximum posteriori framework and a local regression model were proposed respectively to control real-time data-driven character such as human locomotion in [10–12]. Heungil et al. combined the hidden Markov model and automatic encoder to model the underlying functional dynamics inherent in rs-fMRI [13]. At present, the application of CNN on the regression task is very limited. In addition, the long short-term memory network is often used to process time series, for it can establish the correlation between the previous information and the current circumstances [14,15]. To the best of our knowledge, there is no report about combining CNN and LSTM network to predict heating, gas, and electrical loads while considering their correlation.

In this paper, we aim to forecast heating, gas, and electrical loads by combining CNN and LSTM network. Firstly, the Pearson correlation coefficient will be utilized to analyze the temporal correlation between historical loads and current loads, which give the reason for using the LSTM network. Then, a deep learning method composed of CNN and LSTM network could be designed. In addition, the dropout layer is proposed to handle the over-fitting. Finally, the real-world data of CCHP system is used to test the performance of our proposed approaches.

The rest of this paper is organized as follows. Section 2 provides the background of load forecasting. Section 3 analyzes the temporal correlation of the three loads and the coupling between them, and then explains why LSTM should be added to the proposed network. Section 4 introduces the Conv1D, MaxPooling1D, dropout and LSTM layers for load forecasting. Section 5 tests the performance of our proposed approaches and analyses results. Section 6 summaries the conclusions.

2. Literature Review

Heating, gas, and electrical loads forecasting are essential to CCHP systems planning and operations. In respect of time horizons, the loads forecasting can be roughly split into long-term load forecasting, medium-term load forecasting, short-term load forecasting, and very short-term load forecasting, among which the predicted time horizon cut-offs are years, months, hours, and minutes, respectively. This section will provide a brief review of short-term load forecasting.

In the previous literature, several forecasting approaches were proposed for predicting heating, gas, and electrical loads. The conventional methods mainly include autoregressive integrated moving average (ARIMA), model support vector machine (SVM), regression analysis, grey theory (i.e., GM (1,1)) and artificial neural network (ANN).

The current state of heating, gas, and electrical loads is not only related to the surrounding environmental factors but also influenced by past events. The ARIMA and GM (1,1) models predict current load according to historical time series, which can fully consider the trend and transient state. However, they ignore environmental factors. Therefore, when the surrounding environment changes dramatically, the historical trend of the load is not smooth, and the error of these methods may become very large [16,17].

Regression analysis fits the given mathematical formula based on historical data, but it has the drawback that the relationship between loads and features is difficult to be accurately described by a mathematical formula [18]. In the field of computer science, SVM is a supervised learning model, which is often utilized for the task of classification and regression analysis. It is good at solving a large number of complex problems, such as nonlinear, over-fitting, high dimension, and local minimum point. However, the SVM has a slow speed of training large-scale samples [19,20]. As a "black-box" that relies on data and prior knowledge, the traditional ANN can fit complex nonlinear relationships, whereas the traditional ANN also has defects of over-fitting and easy to fall into local optimum [21,22]. In addition, the above methods only account for the impact of the environmental factors on the current loads, ignoring the role of past events.

Recently, the deep learning network has been applied to forecast heating, gas and electrical loads. The deep belief network is designed to forecast day-ahead electricity consumption in [23]. The study cases show that the proposed approach is suitable for short-term electrical load forecasting. In addition, it offers better results than traditional methods. Indeed, the LSTM is good at dealing with time series with long time spans, which is suitable for forecasting short-time loads. Kuan Lu et al. proposed a concatenated LSTM architecture for forecasting heating loads [24]. In order to solve the forecasting problem for the strong fluctuating household load, Weicong Kong et al. improved the household prediction framework with automatic hyper parameter tuning based on LSTM network [14]. The CNN is a neural network designed to process input data that has an intrinsic relationship. Generally, the input data to CNN will have a natural structure to it such that nearby entries are correlated [25,26]. For example, this type of data includes 1-D load time series and 2-D images. The current research mainly focuses on 2-D image recognition. The literature about using CNN to extract the features of time series for forecasting loads is relatively limited. In order to improve the performance of the network, researchers try to combine CNN with LSTM to form a hybrid network. A CNN-LSTM neural network is proposed to extract temporal and spatial features to improve the forecasting accuracy of household load in [27]. Jianfeng et al., designed a hybrid network consisting of the CNN and LSTM to improve the performance of recognizing speech emotion [28]. Similarly, The CNN and LSTM are utilized to automatically detect diabetes in [29]. At present, there is no report on the use of hybrid network consisting of the CNN and LSTM to predict heating, gas, and electrical loads while considering the correlation of these three loads for integrated energy systems.

In addition, previous studies show that the performance of multi-layer is better than that of single-layer for all of the above deep learning models. However, some scholars have found that over-fitting occurs as the number of layers increases. [30,31]. Therefore, it is necessary to find a way that can increase the number of layers without over-fitting.

Taking the above analysis into consideration, it is clear that though the predecessors have made great achievements in heating, gas and electrical loads forecasting, there are still some problems to be solved. For example, how to combine CNN and LSTM to design a hybrid network, which can not only extract the inherent features of the input but also consider the temporal correlation of loads? How to solve the over-fitting? How does the coupling between heating, gas, and electrical loads affect the forecasting results?

To solve these problems for heating, gas, and electrical loads forecasting, a new framework based on deep learning is proposed. The key contributions of this paper can be summarized as follows:

(1) The heating, gas, and electrical loads of the CCHP system are highly coupled. Although there is a lot of literature focusing on load forecasting, the prediction of multiple loads considering their coupling has not been found in the literature. This is the first time to design a network to forecast loads, considering the coupling between them.

(2) Pearson correlation coefficient will be utilized to measure the temporal correlation between historical loads and current loads, to give the reason for using the LSTM network.

(3) The Conv1D layer and MaxPooling1D layer are utilized to inherent features that affect heating, gas, and electrical loads. To prevent over-fitting, the dropout is added between LSTM layers. The LSTM network which could take the influence of previous information into account is adopted to forecast these loads.

3. Analysis of Temporal Correlation

As we all know, loads have temporal correlations, especially electrical loads. For example, if the air conditioner is turned on at the moment, the air conditioning load will continue for some time in the future. Furthermore, there are many methods, such as the GM (1,1) model, that predict next loads based on the trend of historical load series. In the past, heating, gas, and electrical loads systems operated independently and their coupling was not strong. Therefore, few people study the temporal correlation between multiple loads. The heating, gas, and electrical loads can be converted in real-time through related devices in CCHP systems, which lead to a strong temporal correlation of these three loads.

Pearson correlation coefficient whose value ranges from −1 to +1 is able to measure the linear correlation of two variables. In this paper, the Pearson coefficient will be utilized to evaluate the temporal correlation of these three loads. The Pearson correlation coefficient can be expressed as follows [32]:

$$r_{xy} = \frac{\sum\limits_{i=1}^{n} (x_i - \overline{x})(y_i - \overline{y})}{\sqrt{\sum\limits_{i=1}^{n} (x_i - \overline{x})^2} \sqrt{\sum\limits_{i=1}^{n} (y_i - \overline{y})^2}} \tag{1}$$

where $\overline{x}$ stands for the mean of x and $\overline{y}$ stands for the mean of y.

In this study, the dataset comes from a hospital in Beijing, China, which contains hourly data from 1 January, 2015 to 31 December, 2015. The main features include environmental factors, such as moisture content, humidifying capacity, dry bulb temperature, and total radiation. The Pearson coefficient is used to analyze the relationship between current heating, gas, and electrical load (loads at time t) and their historical loads (loads from t-24 to t-1). The results are shown in Figure 1.

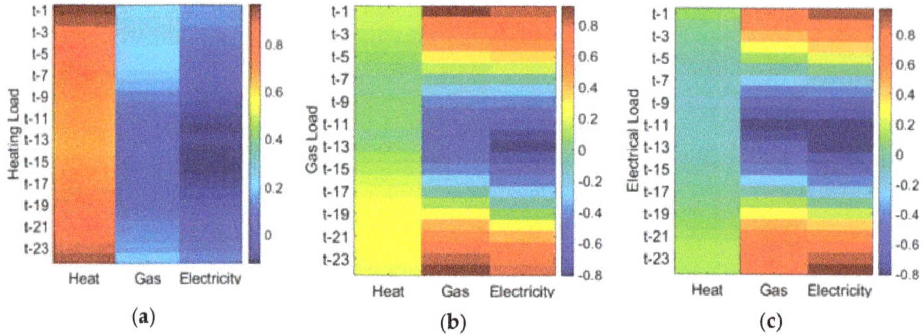

Figure 1. The temporal correlation of heating, gas and electrical loads. (a) Heating load, (b) gas load, (c) electrical load.

On one hand, the Pearson coefficient between the current heating loads and the historical heating loads is large, i.e., the heating load itself has a strong temporal correlation. In addition, the Pearson coefficients between the heating loads and the electrical loads and the gas loads are small, which indicates that there is weak coupling between heating loads and the other two kinds of loads. On the other hand, both the gas load and the electrical load have strong coupling with themselves. Besides, there is a strong coupling between the current gas load and the historical electrical load which ranges from t-1 to t-5. Electrical loads also have a similar conclusion that there is a strong coupling between the current electrical load and the historical gas load which ranges from t-1 to t-4.

As can be seen from the above simulation, the heating, gas, and electrical loads have strong temporal correlation and coupling, which requires the deep learning network to consider these factors.

4. Deep Learning Framework for Forecast Short-Term Loads

4.1. Conv1D Layer and MaxPooling1D Layer

CNN is a neural network designed for processing input data that has an intrinsic relationship. For example, a time series can be thought of as a one-dimensional grid sampled at fixed time intervals, and image data can be viewed as a two-dimensional grid of pixels [33]. CNN has been widely used in image recognition tasks with good performance. As the name implies, the main mathematical operation of convolution neural networks is convolution that is a special linear operation. The matrix multiplication is replaced by convolution layers in CNN.

As is known to all, the convolution is a mathematical operation on two functions of a real-valued argument. The convolution operation can be described as follows:

$$s = x * w \tag{2}$$

where w stands for the weighting function which is called kernel in CNN. x stands for the input function. The output of convolution can be marked as s, which will be called the feature map. $*$ represents the operation of convolution.

In practical problems such as load forecasting, the data of input is a multiple dimensional vector, and the kernel is also a multiply dimensional vector of parameters which are determined by learning method. In this case, the operation of convolution will be applied to multiple dimensions since the kernels and inputs are multiple dimensional. Therefore, the operation of convolution for two-dimensional inputs can be described as follows:

$$s(i, j) = (I * K)(i, j) = \sum_l \sum_m I(l, m) K(i + l, j + m) \tag{3}$$

where I is the two-dimensional data of input, and K is the two-dimensional kernel. S represents the feature map after the operation of convolution.

As shown in Figure 2, a typical CNN consists of a set of layers. The input layer is composed of environmental factors and historical loads. Assuming that the dimension of the input layer is 28, five feature maps are generated after convolution operation. The pooling layers are often inserted between the Conv1D layers. It effectively alleviates over-fitting by reducing the parameters between layers. According to the conclusion from the literature [24], the computationally efficient max pooling showed better results than other candidates, including average pooling and min pooling. The MaxPooling1D layer resizes it spatially and operates on every depth slice of the data.

Generally speaking, the neural network includes one or more Conv1D and MaxPooling1D layers. After extracting features by using Conv1D and MaxPooling1D layers, the outputs will be sent to LSTM layers.

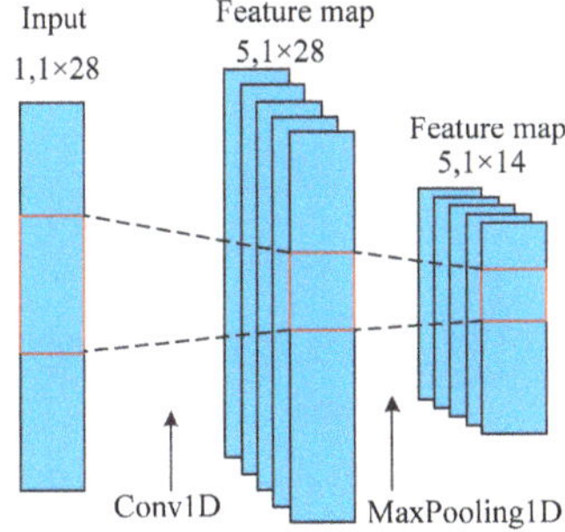

Figure 2. The structure of convolutional neural network.

4.2. LSTM Layer

The recurrent neural network (RNN) is a typical artificial neural network that establishes the temporal correlations between the current circumstances and previous information [34]. Unlike traditional feed forward neural network, the RNN can use their internal memory to process time series of input data. Such characteristic of RNN makes it applicable to load forecasting, because the heating, gas, and electrical loads are affected by environmental features and historical loads.

The common training approaches for RNN mainly include real-time recurrent learning (RTTL) and back propagation through time (BPTT). Compared with RTRL, The BPTT algorithm has a shorter computation time [35]. Therefore, BPTT is often used to train RNN. Because the problems of gradient vanishing and gradient exploding, learning long-range dependencies with RNN is difficult. These problems limit the ability to learn temporal correlations of long-term time series. The long short-term memory (LSTM) was proposed by Hochreiter to solve these problems in 1997 [36]. Broadly speaking, LSTM is one of the RNNs. It not only has memory and forgetting patterns to learn the features of time series flexibly, but also solves the problem of gradient exploding and gradient vanishing. Recently, LSTM networks have achieved great success in numerous sequence prediction tasks, which include speech prediction, handwritten text prediction, etc. Figure 3 shows the block structure of LSTM at a single time step.

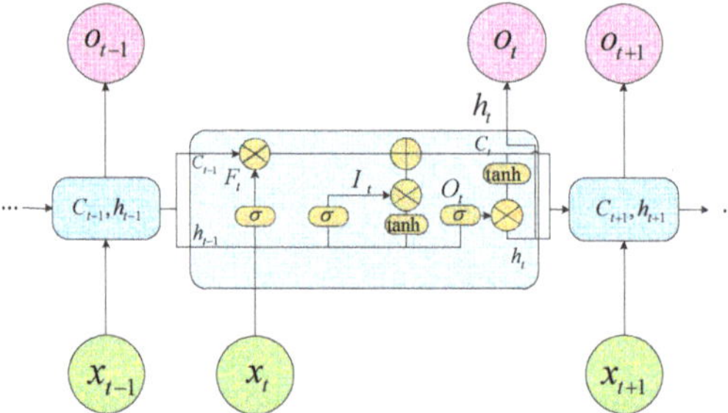

Figure 3. The block structure of long short-term memory (LSTM).

The cell state vector c_t is read and modified through the control of forget gate f_t, input gate i_t and output gates o_t during the whole life cycle, which is the most important structure of the LSTM layer. The current cell state vector c_t will be determined by operating the output vector h_{t-1}, input vector x_t and previous cell state vector c_{t-1} according to the present time steps and the outputs of the previous time step. The formula for the relationship between the variables is as follows:

$$f_t = \sigma_g(W_f x_t + U_f h_{t-1} + b_f) \tag{4}$$

$$i_t = \sigma_g(W_i x_t + U_i h_{t-1} + b_i) \tag{5}$$

$$o_t = \sigma_g(W_o x_t + U_o h_{t-1} + b_o) \tag{6}$$

$$c_t = f_t \circ c_{t-1} + i_t \circ \sigma_c(W_c x_t + U_c h_{t-1} + b_c) \tag{7}$$

$$h_t = o_t \circ \sigma_c(c_t) \tag{8}$$

where $W \in R^{n \times d}$ are the weight matrices. $U \in R^{n \times n}$ are bias vector parameters. The superscripts n is the number of hidden units and d is the number of input features. σ_c is hyperbolic tangent functions and σ_g is the sigmoid function.

The hyperparameter of the hidden unit n should be specified to train the LSTM network. Therefore, the output vector h_t and cell state vector c_t are n-dimensional vectors, which are equal to 0 at the initial time. The LSTM has three sigmoid functions whose output data range from 0 to 1. They are usually regarded as "soft" switches to determine which data should pass through the gate. The signal will be blocked by the gate when the gate is equal to 0. The states of input gate i_t, output gate o_t and forget gate f_t all rely on previous output h_{t-1} and the current input x_t. The signal of forget gate determines what to forget of the previous state c_{t-1}, and the input gate decides what will be preserved in the internal state c_t. After updating the internal state, the output data of LSTM will be determined by the internal state. Similarly, this process will be repeated for the next time steps. In general, the LSTM output of the next time steps can be affected by the information of the previous time steps through this block structure of LSTM.

4.3. Dropout Layer

Previous studies have shown that increasing the number of layers in the neural network does not effectively improve forecasting accuracy. The number of internal parameters of the network increases exponentially when the number of network layers increases. It is prone to over-fitting. After training the network, the network will be created perfectly, but just for the training set. Dropout is a technique that addresses over-fitting [37,38]. As shown in Figure 4, some units are selected randomly and their incoming and outgoing connections are discarded from the network. At each training phase, each unit "exits" the network with a probability p to reduce the parameters of the network. Only the reduced network will be trained in the stage, and the removed units will be reinserted into the network with their original weights.

The probability of discarding hidden units is set to 0.5. In term of input units, the probability should be much lower because if the input units are ignored, the information will be lost directly. By avoiding training all units of the network, the dropout layer can decrease over-fitting. Especially for deep neural networks, dropout technique can significantly shorten the training time.

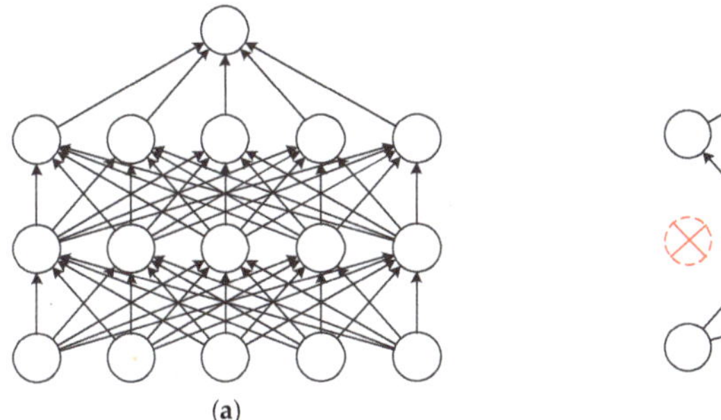
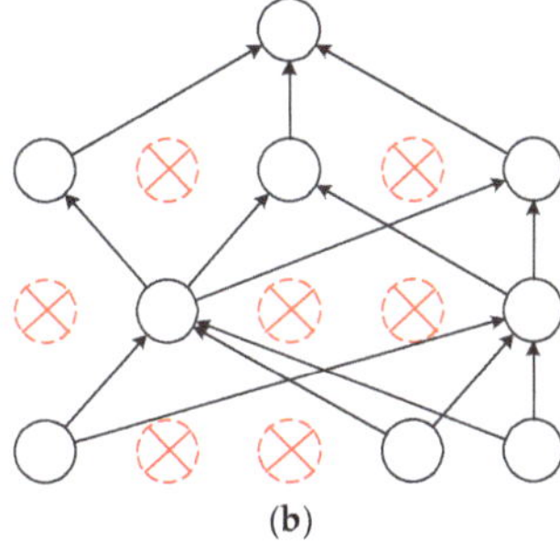

(a)   (b)

Figure 4. Dropout neural network. (**a**) A classical neural net with two hidden layers. (**b**) An example of sparse networks with dropout on the left.

4.4. Framework for Multiple Loads Forecasting Based Deep Learning

Figure 5 shows the framework of short-term loads forecasting based on deep learning. The process of load forecasting is as follows:

(1) The input data include historical loads and environmental factors such as moisture content, humidifying capacity, dry bulb temperature, and total radiation. The min-max normalization is used to bring all input data into the range from 0 to 1.

(2) Next step is to determine the structure of network and parameters, such as the number of LSTM layer, the number of unit in each LSTM layer, the number of CNN layer, the size of kernel weight, the size of pooling, epochs and the size of each batch.

(3) The input data will be sent to Conv1D layers. The MaxPooling1D layer is added between the two Conv1D layers. It extracts the maximum value of the filters and provides useful features while reducing computational cost thanks to data reduction.

(4)   In the LSTM layer, the time steps are sent to relevant LSTM block. The number of LSTM layers can be revised arbitrarily because of the sequential character of the output of the LSTM layer.

The output data of the LSTM layer are used as input of the full connection layer, and the predicted load is output by the full connection layer.

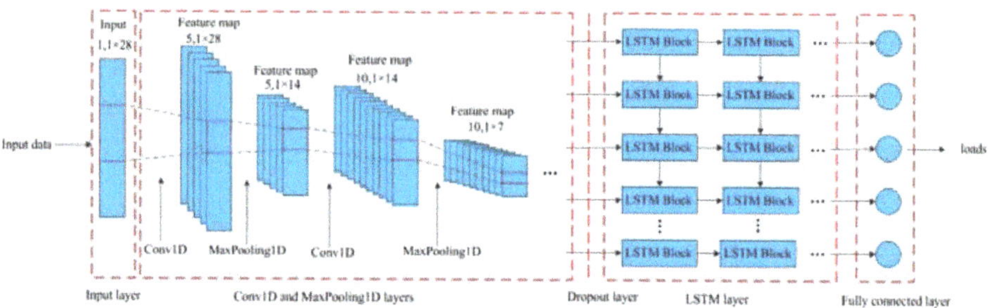

Figure 5. The framework of short-term loads forecasting based on deep learning.

After designing the structure of the neural network, it is necessary to determine the training method. Now, the main training methods of recurrent neural networks, such as LSTM, include real-time recurrent learning (RTRL) and back propagation through time (BPTT). Compared with BPTT, RTRL has lower computational efficiency and longer computing time [33]. Hence, the proposed network will be trained by BPTT. Moreover, previous research suggests Adam approach can achieve better performance than other optimizers, such as Adagrad, Adadelta, RMSProp, and SGD [34]. Therefore, the optimizer for the training proposed approach is Adam. The loss function is MAE.

The main steps of the proposed method can be summarized as follows: (1) Define the CNN-LSTM network, (2) compile the CNN-LSTM network, (3) fit the CNN-LSTM network, (4) predict the loads. The part of the code for the proposed method is shown in Table 1.

Table 1. The code for the proposed method.

Program: A part of codes for building the CNN-LSTM network

```
#1 Define the CNN-LSTM Network
model = Sequential()
model.add(Conv1D(filters=10, kernel_size=3, padding='same', strides=1, activation='relu',input_shape=(1,
Input_num)));
model.add(MaxPooling1D(pool_size=2))
model.add(Dropout(rate=0.25))
model.add(Conv1D(filters=20, kernel_size=3, padding='same', strides=1, activation='relu'))
model.add(MaxPooling1D(pool_size=2))
model.add(Dropout(rate=0.25))
model.add(LSTM(units=24,return_sequences=True))
model.add(LSTM(units=16,return_sequences=True))
model.add(LSTM(units=32,return_sequences=True))
model.add(LSTM(units=16,return_sequences=True))
model.add(LSTM(units=16,return_sequences=True))
model.add(LSTM(units=16))
model.add(Dense(units=1, kernel_initializer='normal',activation='sigmoid'))
#2 Compile the CNN-LSTM network
model.compile(loss='mae', optimizer='adam')
#3 Fit the CNN-LSTM network
history = model.fit(trainX,trainY, epochs=100, batch_size=50,validation_data=(valid3DX, validY), verbose=2,
shuffle=False)
#4 Predict the loads
Predicted_Load = model.predict(testX)
```

4.5. Indicators for Evaluating Result

To measure the predictive effect from various perspectives, mean absolute percentage error (MAPE) will be adopted in this paper. The mathematical formula is as follows:

$$MAPE = \frac{1}{n}\sum_{i=1}^{n}\left|\frac{\hat{y}_i - y_i}{y_i}\right| \tag{9}$$

where n stands for the number of test sets. $\hat{y}_i$ is the forecasting load and y_i is the real load.

5. Case Study

5.1. Experimental Environment and Parameters

The dataset comes from a hospital in Beijing, China, which contains 8760 samples from 1 January 2015 to 31 December 2015. The sample interval was one hour. The loads and corresponding features from 1 January 2015 to 19 October 2015 were used for the training set and the data from 20 October 2015 to 25 November 2015 were used for the validation set. The other data were considered as testing data. The equipment of the integrated energy system mainly included gas boiler, gas-combustion generator, waste-heat recovery system, electric refrigeration unit, lithium bromide refrigeration unit, storage battery and heat storage system. All the proposed methods were conducted using Keras on a notebook computer equipped with Intel (R) Core (TM) i5-6500 CPU @ 3.20 GHz processor and 8 GB of RAM.

In order to verify the validity of the proposed algorithm, the proposed algorithm was compared with the traditional methods (BP network, ARIMA, SVM, LSTM, CNN). The parameters of each algorithm were tested several times in order to achieve optimal performance. However, not all results will be shown here. After many trials, the optimal structure and parameters of each algorithm arweree set as follows:

BP network: The epochs were set to 100. The middle layer consisted of two fully connected layers with 10 and 15 neurons respectively.

ARIMA: The degree of difference was two and the number of autoregressive terms was four. The number of lagged forecast errors was four.

SVM: The kernel function of SVM used the radial basis function (RBF).

LSTM: The neurons' number in the input layer equaled the number of features, and the neurons' number in the output layer was 1. After many trials, the best choice was to use six LSTM layers. The neurons' number in each layer was 32, 16, 32, 16, 16, and 8, respectively.

CNN: After many trials, the best solution of CNN was to use two Conv1D layer and MaxPooling1D layer. The filters were 10 and kernel size was three in the first Conv1D layer. The filters were 20 and kernel size was three in the second Conv1D layer. Both pool sizes of MaxPooling1D were equal to two.

CNN-LSTM: After many trials, the best solution of CNN was to use two Conv1D layer and MaxPooling1D layer. The filters were 10 and kernel size was three in the first Conv1D layer. The filters were 20 and kernel size was threw in the second Conv1D layer. Both pool sizes of MaxPooling1D were equal to two. The best choice was to use six LSTM layers. The neurons' number in each layer was 24, 16, 32, 16, 16, and 16, respectively. Both rates of the dropout were set to 0.25.

This section mainly consists of the following four points: (1) The performance for forecasting heating, gas, and electrical loads was tested in different time steps, (2) the influence of the coupling of heating, gas, and electrical loads on the accuracy of prediction was analyzed, (3) the relationship between the forecasting results and the layers' number of the network is explored, and the influence of the dropout layer on the forecasting accuracy were analyzed, (4) the performance of proposed approaches is compared to traditional methods to validate the efficacy.

5.2. Performance in Different Time Steps

The LSTM network forecasts the loads by using the environmental factors and historical load series whose length can be changed arbitrarily theoretically. If the time steps of power load are too short, it may lead to the insufficiency of learning historical trend. In contrast, if the time steps of power load are too long, it may aggravate the complexity of the proposed methods, which may make the accuracy worse.

To explore how many historical load series are applied to LSTM network for forecasting loads, multiple cases with different time steps which range from 0 to 10 were tested. The average MAPE was calculated by testing the data set 50 times independently. Figures 6–8 show the result of MAPE in different time steps.

As the time steps increase, the overall trend of the heating load of MAPE decreases. This phenomenon suggests that there is a strong temporal correlation between the current heat load and the historical heat load from t-1 to t-10, which is consistent with the conclusions drawn in Figure 1 above. The current gas and electrical loads also have a strong temporal correlation with historical loads from t-1 to t-2, and a weak temporal with historical loads from t-3 to t-10. In this data set, two look-back time steps can achieve the best accuracy for forecasting gas and electrical loads.

When time steps are equal to 0, the MAPE of the heating, gas, and electrical loads are equal to 0.145, 0.158, and 0.143, respectively. In general, considering historical load series can significantly reduce the error for predicting heating, gas, and electrical loads. It shows the need to find a network that can account for temporal correlations to predict heating, gas, and electrical loads, which explains why the LSTM layer is used in the proposed approach.

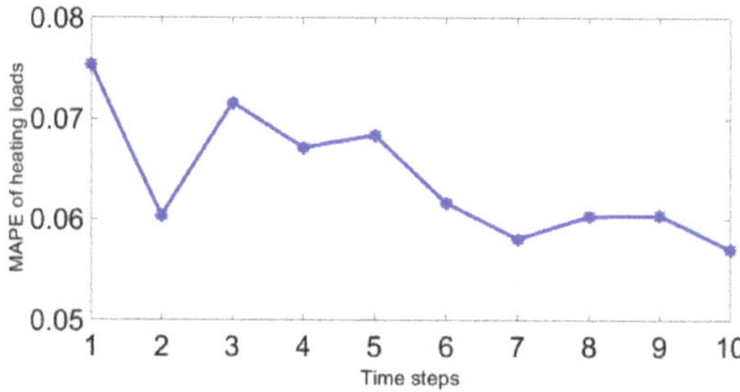

Figure 6. Mean absolute percentage error (MAPE) of heating loads in different time steps.

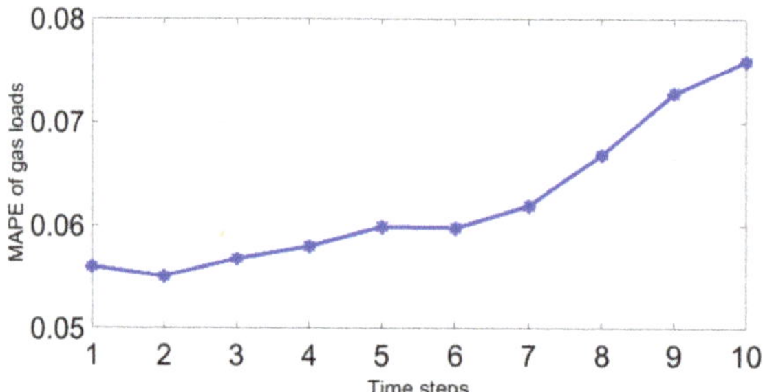

Figure 7. MAPE of gas loads in different time steps.

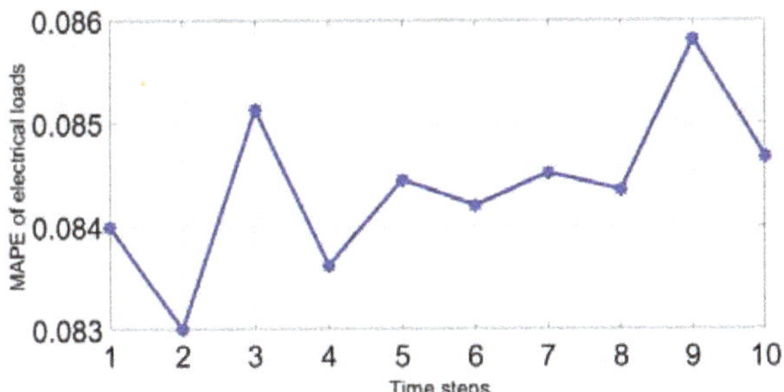

Figure 8. MAPE of electrical loads in time steps.

5.3. The Influence of the Coupling of Heating, Gas and Electrical Loads on the Results

To analyze the impact of the coupling of heating, gas and electrical loads on the forecasting results, eight cases, as shown in Table 2, were designed for simulation. Each case ran 50 times independently to obtain the average MAPE, and the result is shown in Tables 3–5.

The results in Tables 3–5 indicate that:

(1) In terms of heating loads, it is obvious that the forecasting accuracy of Case 2 is higher than that of Case 1, which reveals that the heating loads have a strong temporal correlation and considering temporal correlation helps improve the accuracy of the prediction. By comparing the MAPE of Case 1 and Case 2, it is found that taking the gas load as input will reduce the accuracy for forecasting heating load. Similarly, the conclusion is the same for heating load and electrical load. This is because the coupling between the heating load and the other loads is weak, which is consistent with the conclusions of the previous analysis. The addition of the gas load and the electrical load will interfere with features of the input, which makes the accuracy of the prediction worse.

(2) As far as gas load is concerned, it can be found from Case 3 that the gas load also has a strong temporal correlation. Compared with Case 3 and Case 5, it is evident that the correlation between gas load and heating load is very weak. The input of heating load will lead to a decrease in forecasting accuracy of gas load. On the contrary, the coupling between gas load and electrical load is very strong. Adding electrical load as input is helpful to improve accuracy.

(3) The result of Case 4 from Table 4 shows that there is a strong temporal of electrical load. The input of heating load will lead to a decrease in forecasting accuracy of electrical load. The addition of gas load helps to improve the forecasting accuracy of the electrical load. In general, the best input for forecasting electrical load includes environmental factors, gas and electrical loads.

Table 2. Different case for simulation.

Scenes	Input of Network
Case 1	Environmental features
Case 2	Environmental Feature, heating loads
Case 3	Environmental features, gas loads
Case 4	Environmental features, electrical loads
Case 5	Environmental feature heating and gas loads
Case 6	Environmental features heating and electrical loads
Case 7	Environmental features gas and electrical loads
Case 8	Environmental features heating, gas and electrical loads

Table 3. The average MAPE of heating loads.

Scenes	Case 1	Case 2	Case 5	Case 6	Case 8
MAPE	0.145	0.057	0.065	0.062	0.073

Table 4. The average MAPE of gas loads.

Scenes	Case 1	Case 3	Case 5	Case 7	Case 8
MAPE	0.158	0.060	0.772	0.055	0.064

Table 5. The average MAPE of electrical loads.

Scenes	Case 1	Case 4	Case 6	Case 7	Case 8
MAPE	0.143	0.086	0.092	0.083	0.085

5.4. The Performance of the Dropout Layers

To analyze the relationship between forecasting results and network depth, a sensitive analysis was performed. The number of LSTM layers was increased in turn and the other parameters were kept consistent. Each case ran 50 times independently to obtain the average MAPE, and the results are shown in Figures 9–11.

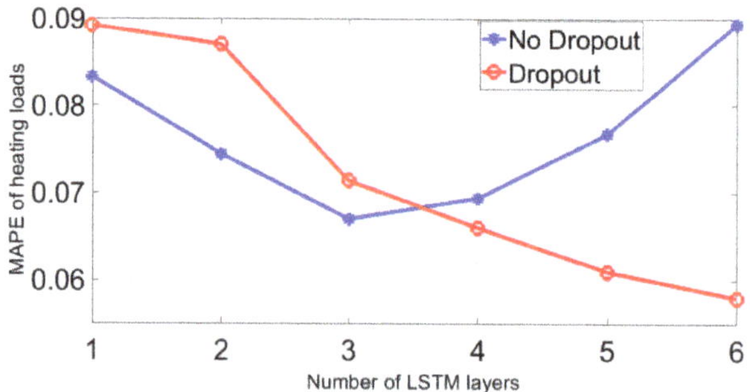

Figure 9. MAPE of heating loads at different network depth.

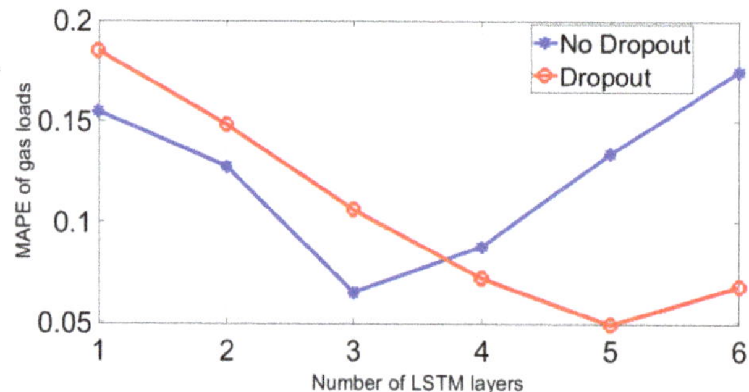

Figure 10. MAPE of gas loads at different network depth.

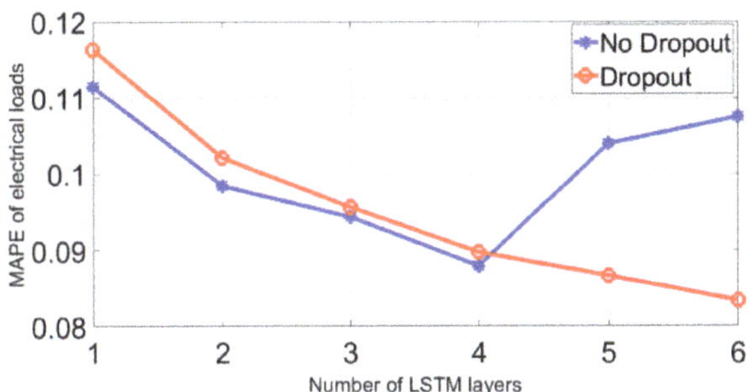

Figure 11. MAPE of electrical loads at different network depth.

As can be seen from the above figures, if there is no dropout layer, the network will achieve the best performance when the number of LSTM layers is three or four. If the number of LSTM layers is further increased, there will be an over-fitting and the MAPE of the loads decreases as the number of LSTM layers increases. The reason for over-fitting is that parameter redundancy of the network rises as the number of LSTM layers increases. In addition, the lack of data diversity also leads to over-fitting.

To tackle the phenomenon of over-fitting, the proposed dropout layers were inlaid to the network. Obviously, when the dropout layers were inlaid to the network, the MAPE of the heating loads and the electrical loads decreased further with the increase of the LSTM layers, which indicates that the dropout layer has the effect of avoiding the over-fitting. Unfortunately, Figure 10 shows that the effect of dropout layers is limited, and it does not completely solve the phenomenon of over-fitting.

5.5. Benchmarking of Short-Term Load Forecasting Methods

To validate the effectiveness of the proposed CNN-LSTM, five loads forecasting methods, including BP network, SVM, ARIMA, CNN, and LSTM were taken as a comparison and assessed under preceding mentioned benchmark (MAPE). Each method ran 50 times independently to obtain the average MAPE of the test set, and the results are shown in Figures 12–14 and Table 6.

Table 6. The MAPE of different algorithms.

Algorithms	Heating Load	Gas Load	Electrical Load
BP	0.067	0.064	0.099
SVM	0.065	0.062	0.096
ARIMA	0.071	0.067	0.131
CNN	0.062	0.060	0.092
LSTM	0.060	0.057	0.088
CNN-LSTM	0.056	0.055	0.082

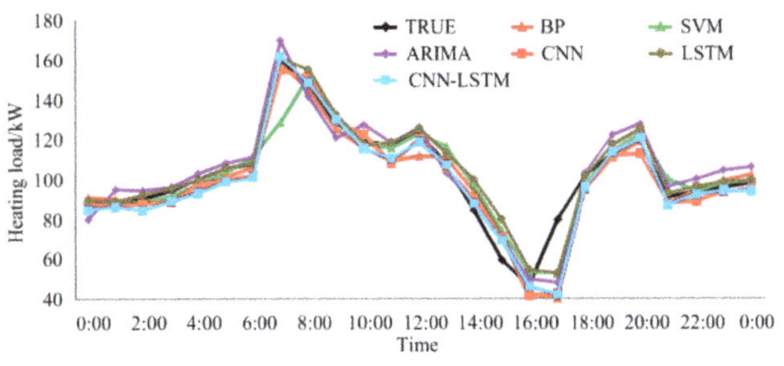

Figure 12. The unfolded topological graph of heating loads.

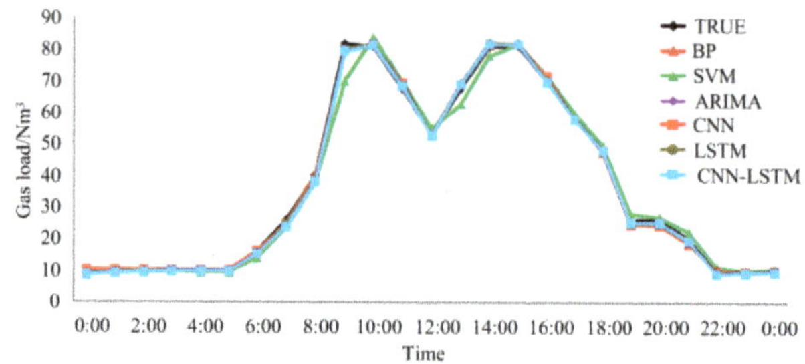

Figure 13. The unfolded topological graph of gas loads.

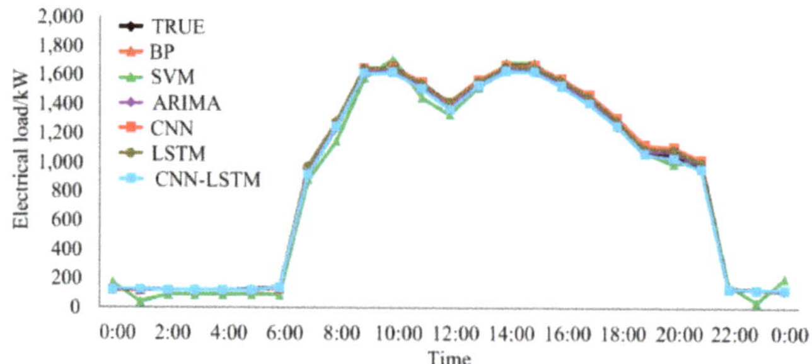

Figure 14. The unfolded topological graph of electrical loads.

The results in figures and Table 6 indicate that:

(1)   The MAPE of the electrical load is greater than that of the heating and gas load, implying that the electrical load has relatively strong volatility compared with other loads. The heating and gas load are more regular and easier to predict. In this data set, the MAPE of heating, gas, and electrical load is 5.6%, 5.5%, and 8.2%, respectively.

(2)   ARIMA has the worst performance because it predicts the load based on the trend of the historical series, without considering the influence of environmental factors. Especially when the environment changes drastically, the forecasting accuracy at the inflection point is very poor. The forecasting accuracy of BP network and SVM is low because of the limitations of their models that make it impossible to pre-learn complex data through unsupervised training. Compared with the deep learning network such as CNN and LSTM, the performance of BP network and SVM is relatively poor. CNN can effectively extract the characteristics of input data, and the forecasting accuracy is higher than that of BP network and SVM. However, CNN cannot deal with the temporal correlation of heating, gas, and electrical loads, which leads to the limitation of forecasting accuracy. Combining CNN and LSTM to construct a hybrid model, it can not only effectively extract the features of input data, but also take into account the temporal correlation of loads. Compared with other traditional methods, CNN-LSTM has the highest forecasting accuracy.

(3)   Figures 12–14 demonstrate the real loads and forecasted loads by different methods on a random day, 11 December 2015. As shown in the figures, the proposed approach has a good performance at spikes and troughs. Taking heating load as an example, during peak and trough periods, the heating load has strong volatility and uncertainty, which makes traditional algorithms unable to accurately predict the load during these periods. However, the morning peak at 7:00 a.m. and afternoon valley at 4:00 p.m. are accurately captured by the proposed approach, which further reflects the superiority of the proposed approach.

6. Conclusions

This paper tries to explore the performance of the CNN-LSTM network for load forecasting considering the coupling of heating, gas, and electrical loads. A novel dropout layer is proposed to successfully solve the phenomenon of over-fitting due to the lack of data diversity and network parameter redundancy. The proposed approach can not only effectively extract the features of input data, but also take temporal correlation of heating, gas, and electrical loads into account. The case study provides the following conclusions:

(1) For heating, gas, and electrical loads, there is a strong temporal correlation between the current loads and historical loads. The case study shows that considering historical load series can reduce the error for predicting heating, gas, and electrical loads.

(2) The coupling between the heating loads and the other loads is weak. Taking the gas loads and the electrical loads as input will make the accuracy of the heating loads worse. The coupling between gas loads and electrical loads is very strong. Adding electrical load as input is helpful to improve the accuracy of gas loads. Similarly, adding gas loads to input data is helpful to improve the forecasting accuracy of electrical loads.

(3) The dropout layer can avoid over-fitting to a certain extent, as well as improve the accuracy for predicting heating, gas, and electrical loads. The dropout layer cannot completely solve the over-fitting where the number of network layers is too large.

(4) Compared with other algorithms (BP network, SVM, ARIMA, CNN, and LSTM), the proposed approach has higher forecasting accuracy and can accurately predict the load during peak and trough periods.

For future work, we can try to expand the work of this article from the following three directions:

(1) We could try to find a technique that can completely solve the over-fitting.

(2) The other deep learning frameworks such as generative adversarial networks (GAN) [39], restricted Boltzmann machines (RBM) [40–42], hidden Markov models [43], dilated convolutional neural network [44,45] and graphical models [46], are also used to forecast heating, gas, and electrical loads. These frameworks are widely used in image recognition, signal processing, and image generation. How to apply these frameworks to load forecasting needs further research. Generally speaking, the function of CNN is to extract the features of input data. Different tasks can be accomplished by using the extracted features as input data of classifiers, predictors, and generators. For example, the GAN's generator consisting of convolution layers can model the power load profiles, and the GAN's discriminator consisting of convolution layers can classify the power load.

(3) Due to the limitations of the data set, this paper only considers the influence of environmental factors and historical data on prediction accuracy. The multimedia features can be taken into account in the future [47,48].

Author Contributions: Funding acquisition, R.Z.; methodology, W.G.; supervision, R.Z.; validation, X.G.; visualization, X.G.; writing—review & editing, W.G.

Funding: This work was supported by the Supporting Project of Electrical Engineering Laboratory of Key Laboratory of Tibet Agriculture and Animal Husbandry University (Grant No. 2017DQ-ZN-01) and the Key Scientific Research Projects of the Science and Technology Department of Tibet Autonomous Region (Grant No. Z2016D01G01/01).

Conflicts of Interest: The authors declare no conflict of interest.

References

1. Wei, M.; Yuan, W.; Fu, L.; Zhang, S.; Zhao, X. Summer performance analysis of coal-based cchp with new configurations comparing with separate system. *Energy* **2018**, *143*, 104–113. [CrossRef]

2. Wu, J.; Wang, J.; Wu, J.; Ma, C. Exergy and exergoeconomic analysis of a combined cooling, heating, and power system based on solar thermal biomass gasification. *Energies* **2019**, *12*, 2418. [CrossRef]

3. Wegener, M.; Isalgué, A.; Malmquist, A.; Martin, A. 3e-analysis of a bio-solar cchp system for the andaman islands, india—A case study. *Energies* **2019**, *12*, 1113. [CrossRef]

4. Zheng, X.; Wu, G.; Qiu, Y.; Zhan, X.; Shah, N.; Li, N.; Zhao, Y. A minlp multi-objective optimization model for operational planning of a case study cchp system in urban china. *Appl. Energy* **2018**, *210*, 1126–1140. [CrossRef]

5. Wu, B.; Li, K.; Ge, F.; Huang, Z.; Yang, M.; Siniscalchi, S.M.; Lee, C. An end-to-end deep learning approach to simultaneous speech dereverberation and acoustic modeling for robust speech recognition. *IEEE J. Sel. Top. Signal Process.* **2017**, *11*, 1289–1300. [CrossRef]

6. Heo, Y.J.; Kim, S.J.; Kim, D.; Lee, K.; Chung, W.K. Super-high-purity seed sorter using low-latency image-recognition based on deep learning. *IEEE Robot. Autom. Lett.* **2018**, *3*, 3035–3042. [CrossRef]

7. Yan, H.; Wan, J.; Zhang, C.; Tang, S.; Hua, Q.; Wang, Z. Industrial big data analytics for prediction of remaining useful life based on deep learning. *IEEE Access* **2018**, *6*, 17190–17197. [CrossRef]

8. Rachmadi, M.F.; Valdés-Hernández, M.D.C.; Agan, M.L.F.; Di Perri, C.; Komura, T. Segmentation of white matter hyperintensities using convolutional neural networks with global spatial information in routine clinical brain MRI with none or mild vascular pathology. *Comput. Med. Imaging Graph.* **2018**, *66*, 28–43. [CrossRef]

9. Krizhevsky, A.; Sutskever, I.; Hinton, G.E. ImageNet Classification with Deep Convolutional Neural Networks. *Neural Inf. Process. Syst.* **2012**, *141*, 1097–1105. [CrossRef]

10. Holden, D.; Komura, T.; Saito, J. Phase-functioned neural networks for character control. *ACM Trans. Graph.* **2017**, *36*, 42. [CrossRef]

11. Mousas, C.; Newbury, P.; Anagnostopoulos, C. Evaluating the covariance matrix constraints for data-driven statistical human motion reconstruction. In Proceedings of the Spring Conference on Computer Graphics, New York, NY, USA, 28–30 May 2014; pp. 99–106.

12. Mousas, C.; Newbury, P.; Anagnostopoulos, C.-N. Data-Driven Motion Reconstruction Using Local Regression Models. In Proceedings of the International Conference on Artificial Intelligence Applications and Innovations, Rhodos, Greece, 19–21 September 2014; pp. 364–374.

13. Suk, H.; Wee, C.Y.; Lee, S.W.; Shen, D. State-space model with deep learning for functional dynamics estimation in resting-state fMRI. *NeuroImage* **2016**, *129*, 292–307. [CrossRef] [PubMed]

14. Kong, W.; Dong, Z.Y.; Jia, Y.; Hill, D.J.; Xu, Y.; Zhang, Y. Short-term residential load forecasting based on lstm recurrent neural network. *IEEE Trans. Smart Grid* **2019**, *10*, 841–851. [CrossRef]

15. Khodayar, M.; Wang, J. Spatio-temporal graph deep neural network for short-term wind speed forecasting. *IEEE Trans. Sustain. Energy* **2019**, *10*, 670–681. [CrossRef]

16. Kouzelis, K.; Bak-Jensen, B.; Mahat, P.; Pillai, J.R. In A simplified short term load forecasting method based on sequential patterns. In Proceedings of the IEEE PES Innovative Smart Grid Technologies, Istanbul, Turkey, 12–15 October 2014; pp. 1–5.

17. Wang, Z.-X.; Li, Q.; Pei, L.-L. A seasonal gm(1,1) model for forecasting the electricity consumption of the primary economic sectors. *Energy* **2018**, *154*, 522–534. [CrossRef]

18. Zhao, J.; Liu, X. A hybrid method of dynamic cooling and heating load forecasting for office buildings based on artificial intelligence and regression analysis. *Energy Build.* **2018**, *174*, 293–308. [CrossRef]

19. Barman, M.; Dev Choudhury, N.B.; Sutradhar, S. A regional hybrid goa-svm model based on similar day approach for short-term load forecasting in assam, india. *Energy* **2018**, *145*, 710–720. [CrossRef]

20. Chia, Y.Y.; Lee, L.H.; Shafiabady, N.; Isa, D. A load predictive energy management system for supercapacitor-battery hybrid energy storage system in solar application using the support vector machine. *Appl. Energy* **2015**, *137*, 588–602. [CrossRef]

21. Li, K.; Xie, X.; Xue, W.; Dai, X.; Chen, X.; Yang, X. A hybrid teaching-learning artificial neural network for building electrical energy consumption prediction. *Energy Build.* **2018**, *174*, 323–334. [CrossRef]

22. Singh, P.; Dwivedi, P. Integration of new evolutionary approach with artificial neural network for solving short term load forecast problem. *Appl. Energy* **2018**, *217*, 537–549. [CrossRef]

23. Dedinec, A.; Filiposka, S.; Dedinec, A.; Kocarev, L. Deep belief network based electricity load forecasting: An analysis of macedonian case. *Energy* **2016**, *115*, 1688–1700. [CrossRef]

24. Kuan, L.; Zhenfu, B.; Xin, W.; Xiangrong, M.; Honghai, L.; Wenxue, S.; Zijian, Z.; Zhimin, L. In Short-term chp heat load forecast method based on concatenated lstms. In Proceedings of the 2017 Chinese Automation Congress (CAC), Jinan, China, 20–22 October 2017; pp. 99–103.

25. Zhang, F.; Xi, J.; Langari, R. Real-time energy management strategy based on velocity forecasts using v2v and v2i communications. *IEEE Trans. Intell. Transp. Syst.* **2017**, *18*, 416–430. [CrossRef]

26. Wang, L.; Scott, K.A.; Xu, L.; Clausi, D.A. Sea ice concentration estimation during melt from dual-pol sar scenes using deep convolutional neural networks: A case study. *IEEE Trans. Geosci. Remote Sens.* **2016**, *54*, 4524–4533. [CrossRef]

27. Kim, T.-Y.; Cho, S.-B. Predicting residential energy consumption using CNN-LSTM neural networks. *Energy* **2019**, *182*, 72–81. [CrossRef]

28. Zhao, J.; Mao, X.; Chen, L. Speech emotion recognition using deep 1D & 2D CNN LSTM networks. *Biomed. Signal Process. Control* **2019**, *47*, 312–323.

29. Swapna, G.; Kp, S.; Vinayakumar, R. Automated detection of diabetes using CNN and CNN-LSTM network and heart rate signals. *Procedia Comput. Sci.* **2018**, *132*, 1253–1262.

30. Radford, A.; Metz, L.; Chintala, S. Unsupervised Representation Learning with Deep Convolutional Generative Adversarial Network. In Proceedings of the International Conference on Learning Representations, San Juan, Puerto Rico, 2–4 May 2016; pp. 1–16.

31. Panchal, G.; Ganatra, A.; Shah, P.; Panchal, D. Determination of over-learning and over-fitting problem in back propagation neural network. *Int. J. Soft Comput.* **2011**, *2*, 40–51. [CrossRef]

32. Xu, H.; Deng, Y. Dependent evidence combination based on shearman coefficient and pearson coefficient. *IEEE Access* **2018**, *6*, 11634–11640. [CrossRef]

33. Abdeljaber, O.; Avci, O.; Kiranyaz, S.; Gabbouj, M.; Inman, D.J. Real-time vibration-based structural damage detection using one-dimensional convolutional neural networks. *J. Sound Vib.* **2017**, *388*, 154–170. [CrossRef]

34. Qing, X.; Niu, Y. Hourly day-ahead solar irradiance prediction using weather forecasts by lstm. *Energy* **2018**, *148*, 461–468. [CrossRef]

35. Graves, A.; Schmidhuber, J. Framewise phoneme classification with bidirectional LSTM and other neural network architectures. *Neural Netw.* **2005**, *18*, 602–610. [CrossRef]

36. Hochreiter, S.; Schmidhuber, J. Long short-term memory. *Neural Comput.* **1997**, *9*, 1735–1780. [CrossRef] [PubMed]

37. Sawaguchi, S.; Nishi, H. Slightly-slacked dropout for improving neural network learning on FPGA. *ICT Express* **2018**, *4*, 75–80. [CrossRef]

38. Zhang, Y.-D.; Pan, C.; Sun, J.; Tang, C. Multiple sclerosis identification by convolutional neural network with dropout and parametric ReLU. *J. Comput. Sci.* **2018**, *28*, 1–10. [CrossRef]

39. Rekabdar, B.; Mousas, C.; Gupta, B. Generative Adversarial Network with Policy Gradient for Text Summarization. In Proceedings of the 13th IEEE International Conference on Semantic Computing, Newport Beach, CA, USA, 30 January–1 February 2019; pp. 204–207.

40. Ngiam, J.; Khosla, A.; Kim, M.; Nam, J.; Lee, H.; Ng, A.Y. Multimodal deep learning. In Proceedings of the 28th International Conference on International Conference on Machine Learning, Bellevue, WA, USA, 28 June–2 July 2011; pp. 689–696.

41. Mousas, C.; Anagnostopoulos, C. Learning Motion Features for Example-Based Finger Motion Estimation for Virtual Characters. *3D Res.* **2017**, *8*, 25. [CrossRef]

42. Nam, J.; Herrera, J.; Slaney, M.; Smith, J.O. Learning Sparse Feature Representations for Music Annotation and Retrieval. In Proceedings of the 13th International Society for Music Information Retrieval Conference, Porto, Portugal, 8–12 October 2012; pp. 565–570.

43. Abdelhamid, O.; Mohamed, A.; Jiang, H.; Penn, G. Applying Convolutional Neural Networks concepts to hybrid NN-HMM model for speech recognition. In Proceedings of the 2012 IEEE International Conference on Acoustics, Speech and Signal Processing, Kyoto, Japan, 25–30 March 2012; pp. 4277–4280.

44. Rekabdar, B.; Mousas, C. Dilated Convolutional Neural Network for Predicting Driver's Activity. In Proceedings of the 21st International Conference on Intelligent Transportation Systems, Maui, HI, USA, 4–7 November 2018; pp. 3245–3250.

45. Li, R.; Si, D.; Zeng, T.; Ji, S.; He, J. Deep convolutional neural networks for detecting secondary structures in protein density maps from cryo-electron microscopy. In Proceedings of the IEEE International Conference on Bioinformatics and Biomedicine, Shenzhen, China, 15–18 December 2016; pp. 41–46.

46. Saito, S.; Wei, L.; Hu, L.; Nagano, K.; Li, H. Photorealistic Facial Texture Inference Using Deep Neural Networks. In Proceedings of the 2017 IEEE Conference on Computer Vision and Pattern Recognition, Honolulu, HI, USA, 21–26 July 2017; pp. 2326–2335.

47. Amato, F.; Castiglione, A.; Moscato, V.; Picariello, A.; Sperli, G. Multimedia summarization using social media content. *Multimed. Tools Appl.* **2018**, *77*, 17803–17827. [CrossRef]
48. Hu, D.; Li, J.; Liu, Y.; Li, Y. Flow Adversarial Networks: Flowrate Prediction for Gas-Liquid Multiphase Flows across Different Domains. *IEEE Trans. Neural Netw. Learn. Syst.* **2019**, 1–13. [CrossRef]

Article

Multivariate Analysis to Relate CTOD Values with Material Properties in Steel Welded Joints for the Offshore Wind Power Industry

Álvaro Presno Vélez [1], Antonio Bernardo Sánchez [1,*], Marta Menéndez Fernández [1] and Zulima Fernández Muñiz [2]

[1] Department of Mining Technology, Topography and Structures, University of León, 24071 León, Spain; a.presno.velez@gmail.com (Á.P.V.); marta.menendez@unileon.es (M.M.F.)
[2] Department of Mathematics, University of Oviedo, 33007 Oviedo, Spain; zulima@uniovi.es
* Correspondence: abers@unileon.es; Tel.: +34-987-293-554

Received: 28 September 2019; Accepted: 18 October 2019; Published: 21 October 2019

Abstract: The increasingly mechanical requirements of offshore structures have established the relevance of fracture mechanics-based quality control in welded joints. For this purpose, crack tip opening displacement (CTOD) at a given distance from the crack tip has been considered one of the most suited parameters for modeling and control of crack growth, and it is broadly used at the industrial level. We have modeled, through multivariate analysis techniques, the relationships among CTOD values and other material properties (such as hardness, chemical composition, toughness, and microstructural morphology) in high-thickness offshore steel welded joints. In order to create this model, hundreds of tests were done on 72 real samples, which were welded with a wide range of real industrial parameters. The obtained results were processed and evaluated with different multivariate techniques, and we established the significance of all the chosen explanatory variables and the good predictive capability of the CTOD tests within the limits of the experimental variation. By establishing the use of this model, significant savings can be achieved in the manufacturing of wind generators, as CTOD tests are more expensive and complex than the proposed alternatives. Additionally, this model allows for some technical conclusions.

Keywords: crack tip opening displacement; steels; welded or bonded joints; multivariate regression model; marine structures

---

1. Introduction

As the burgeoning offshore wind power industry grows, so too do the technical demands on the metal frames and primary structures that sustain them. These structures are under enormous dynamic stresses due to the effects of their moving parts, wind, currents, tides, and waves. Within this sector, the quality control of the welded joints of these structures is of the utmost importance, considering that welding defects are widely considered as potential spots for structural failure initiation [1]. The study based on fracture mechanics of parameters such as CTOD, in the context of crack nucleation and fatigue crack growth, has become essential for manufacturers, designers, classification societies, and inspectors. The fatigue life calculation occupies a prominent place in codes, standards, and rules [2–4]. Such fatigue analysis is based on "rule-based" methods or direct calculation based on Stress-Cycles data models, determined by fatigue testing of the considered welded details and linear damage hypothesis.

As this approach is rarely possible (due to the full fatigue test required for the welded details), the fatigue analysis may alternatively be based on fracture mechanics. The classification societies' crack growth models use the classic formulation of the Paris–Erdogan law, with developments for the classical plastic hinge models (firstly developed by the British Standards Institution and published in 1979). According to the vast work of Zhu and Joyce [5], the stress intensity factor K [6], the crack tip opening displacement (CTOD) [7], the J-integral [8], and the crack tip opening angle (CTOA) (developed for thin-walled materials) are the most relevant parameters used in fracture mechanics. Out of these various parameters of the interaction of the materials with the formation and propagation of cracks or defects, the critical crack tip opening displacement (CTOD) at a given distance from the crack tip is the most suited for modeling stable crack growth and instability during the fracture process [9]. Currently, the tests are carried out by discarding the plastic hinge model and adopting the J-conversion, using recognized standards such as the (British Standard) BS-7910, (American Petroleum Institute) API-579, and (American Society for Testing and Materials) ASTM E1290.

CTOD testing requires the preparation of a notch with a specific geometry that promotes the nucleation of a stable and uniform crack in a delimited area [10]. The crack grows under the action of dynamic mechanical forces that are generally transmitted with huge oleo-hydraulic equipment and controlled by precision extensometers. The uncertainty of the test methods, as well as the sensitivity to any internal defect, make it necessary to carry out several of these tests to guarantee representative values.

The CTOD tests are expensive, as they require significant investments in testing machinery, software, expertise, and outsourcing of services [11]. The destruction of large quantities of ad hoc welded material is also required (ASTM E1290-08e1c (2008) [12]). Additionally, deadlines offered by the testing laboratories exceed the average for other quality control tests in welded unions. Considering the case of welded joints, in addition to the properties of the base material, dozens of other variables related to the welding process could affect the features of the final welded material. Therefore, if the CTOD test result does not fulfil the requirements, it is very difficult for technicians to infer which changes in the variables could lead to an improvement of the CTOD results.

The aim of the present work is to evaluate the possibility of using multivariate mathematical models to correlate the CTOD parameter with other test results that are simpler and cheaper to measure, and also well known by the parties involved.

2. Selection of Input Variables and Experimental Phase

The multivariate analysis consists of a series of appropriate statistical methods (such as multiple regression, logistic regression, analysis of variance (ANOVA), or cluster analysis, to name a few) used when numerous observations are performed on the same object in one or several samples. Those methods allow the creation of formal hypothesis tests when given a structure of input–output data. Expressing a variable as a function of a set of underlying intercorrelated variables is among the possible hypotheses [13].

2.1. Selection of Input Variables

The selection of these so-called explanatory variables was done considering the industrial approach of this research work. Among the numerous variables with proven effects on the material properties (see Table 1 for a non-exhaustive selection proposed by Dunne et al. [14] and Haque and Sudhakar [15]), the following ones were selected due to their widespread use in the industry, relatively cheap measurement, and possibility to be determined in modest-quality control laboratories. Also, the chosen variables are part of the testing process required by the design codes, rules, and standards for the design, qualification, and control of welded joints. Therefore, these values are usually available (or easy to gather), there are clear acceptance criteria, and their effects on the CTOD and on welded joints are widely recognized.

Table 1. Non-exhaustive selection of variables with proven influence on material properties.

Variables	
Carbon (wt %)	Plate thickness (mm)
Manganese (wt %)	Post Welding Heat Treatment (PWHT) holding time
Silicon (wt %)	PWH cooling rate/method
Sulphur (wt %)	Test piece orientation
Phosphorus (wt %)	Test temperature
Aluminum (wt %)	Yield Strength (MPa)
Boron (wt %)	Ultimate Tensile Strength (MPa)
Molybdenum (wt %)	Charpy toughness (J)
Oxygen (wt %)	Grain boundaries and orientation
Nitrogen (wt %)	Hardness
V% fraction of reaustenized region	Grain boundary ferrite
V% fraction of double-reheated zone	Intragranular polygonal ferrite
Grain refined subzone	Grain coarsened subzone
Non-metallic inclusions	Mean 3D diameter of inclusions

2.1.1. Microstructure

The microstructure of the material in the area in which the CTOD value is to be determined will be considered one of the input variables. Some authors [16–23] have studied the relation between microstructure characteristics and fracture mechanics properties and supports, and the influence of grain size, angle of grain boundaries, orientation, and inclusions on the nucleation and propagation of cracks. The average size of the metallic grains in the area of interest was determined according to ASTM E112 (2013) [24] (determined by optical microscopy) to represent this variable. The specimens were polished and prepared according to the recommendation of E3-11 (Guide for Preparation of Metallographic Specimens) [25] for Al_2O_3 abrasive (1200 American National Standards Institute grit number), with rotation and etching reagent no. 77 (E407-07 Standard Practice for Microteaching Metals and Alloys) [26].

The limitation in obtaining samples with different surface orientations (see Figure 1) appropriate for eventual non-equiaxed grain shapes was corrected with the implementation of an arbitrary multiplication factor, depending on the grain contour. Any possible heterogeneity in the area of interest is expected to be statistically covered by the experimental design. Having considered the industrial approach, other well-known techniques that require specific equipment, such as scanning electron microscope (SEM), were not used. Also, as failure types are not considered as study variables, the critical grain size for brittle fracture was not considered.

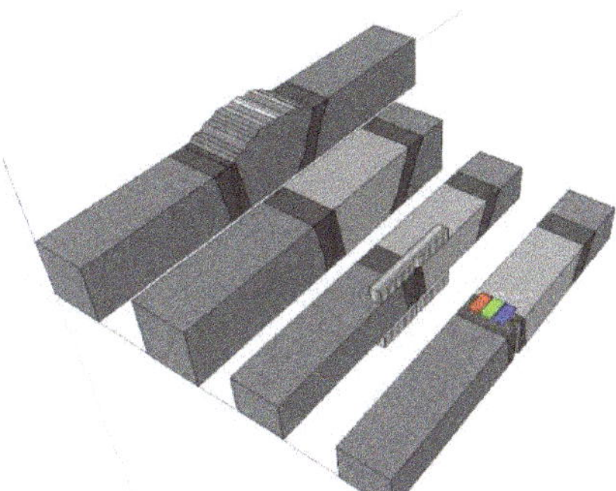

Figure 1. Sampling position. Color zones mark targeted areas for microstructural, hardness, and chemical analysis (red, green and blue).

2.1.2. Chemical Composition

The chemical composition of the material is a well-known factor that exerts influence on the mechanical properties [27–29].

Samples were analyzed by optical emission spectrometry and X-ray diffraction using a Niton $^\circledR$ XL2 analyzer and a Spectromax metal analyzer. Results were statistically processed to offer the best-weighted average estimator considering the different uncertainties of the testing method and for the following elements: C, Mn, Si, Cr, Ni, Mo, and V. Both the test procedure and the uncertainty calculation used were approved by the testing laboratory. For the implementation of the chemical composition into the mathematical model, we considered the influence of the different elements using the carbon equivalent (CE) index, expressed in Equation (1). Among the numerous CE formulae available in the bibliography, we chose American Welding Society (AWS) D1.1 [30], which was cited in [29] and is also known as the International Institute of Welding (IIW) carbon equivalent.

This expression was selected considering its precision for mechanical and microstructural properties [27]:

$$CE_{index} = C + \frac{(Mn + Si)}{6} + \frac{(Cr + Mo + V)}{5} + \frac{(Ni + Cu)}{15} \tag{1}$$

where all values involved represent the mass percentage composition [w/w%]. Therefore, the result is a non-dimensional continuous variable.

2.1.3. Mechanical Strength

The mechanical resistance plays a fundamental role, and forms a constitutive part, in fracture mechanics [31]. Also, the determination and control of its value is a fundamental part of the quality control of the material properties (for structural materials). Tensile test results were discarded due to the impossibility to take measurements exclusively in the small area of interest, as all the subsized specimens proposed by the standards exceed the capability of the testing machine (too small) or destroy valuable testing material (too big). Nevertheless, according to numerous publications (e.g., ASM Handbook for carbon steel), there is a consistent and almost linear relation between ultimate tensile strength (UTS) and hardness. Therefore, hardness measurements according to ASTM E92 (Hardness Vickers 10) (2017) [32] were taken from the samples to estimate the mechanical resistance of the material. Standardized Vickers indenters (Class B) were used with a load of 98.7 N (HV 10) and an optical indentation measurement. The average value of a set of three indentations (considering 2 mm of space between tests) were examined for each sample.

2.1.4. Toughness

Previous studies [15,16,33,34] support the relation between impact testing results (measured as Charpy V-notch (CVN) energy values) and fracture toughness. Some correlations have been adopted by the standards ASME Boiler and Pressure Vessel Code (BPVC) XI (2017) [35] and API 579 [3].

CVN tests, according to ASTM E23 (2018) [36], were performed on the samples. Subsize Charpy simple-beam V-notch impact test specimens were used (2.5 mm, according to Figure A3.1 from ASTM E-23), with the notch aligned with the future CTOD sample notch [10]. All tests were performed at room temperature (between 20 and 25 °C) with a 300 J pendulum device. Three specimens (instead of two) were used for each toughness characterization to ensure representative values (see Figure 1), due to the sample size limitation. Measurement of lateral expansion or the fracture region size was not considered.

2.2. Experimental Design

For a multivariate statistical study (with a suitable uncertainty), it is required to reach a determined critical mass of input data. This number is undetermined, and it will be verified after the modeling [17]. In addition, a wide range that covers the industrial interest is required for the explanatory variables.

It is expected that the heterogeneities on the physical properties of the welded joints and the uncontrolled variables were arbitrarily distributed among the observations, according to the random principle [37]. Nevertheless, the sampling and test position was designed to minimize the effect of these heterogeneities (see Figure 1) by reducing, as much as possible, the area from where the results are obtained.

The first approximation of the complete number of tests was estimated considering the guide of factorial experimental design, computed for four variables, each with two levels, and one replication. A total of 72 complete sets of data were obtained through testing campaigns.

The welded coupons (from where the specimens were extracted for testing) were kindly transferred from manufacturing companies. This guarantees the reproducibility with respect to real welding designs and manufacturing processes, and also the applicability of the ranges used. Nevertheless, it also limits the number of available samples, and as the authors cannot control the range of variability of the study variables, experimental designs with surface analysis or complex factorial designs cannot be used. These limitations of the used experimental set may affect the accuracy of future models (by not gathering the critical amount of data for multivariate models) and prevent the use of more explanatory variables.

2.3. Samples

All the samples were extracted from 36 welded coupons of at least 400 mm length in the direction of the weld. Those coupons were welded for real Welding Procedure Qualification records following real Welding Procedure Specifications, then kindly transferred from manufacturing companies for this project. The thickness of the coupons varied between 20–75 mm and were considered representative of offshore manufacturing. *K* and V-bevels were used, and the base materials and consumables were standard within the manufacturing sector. Several different consumables and four base materials classified according to EN 10225 [38] (low-alloy steels S355 G5 + M, S355 G10 + M, S420 G2 + M, and S460G2 + M together) were used, together with two structural steels from EN 10025:2010 [39] (low-carbon steels S275 J2 and S355 K2). The following table (Table 2) summarizes the range of different relevant variables during the welding process that may have an influence on the properties of the welding coupons. These ranges are considered as representative of the structural welding processes of the offshore wind power industry.

Table 2. The range for different variables of the test coupons.

Variables	Min.	Max.	Variables	Min.	Max.
Wire diameter	1.2 mm	2.8 mm	Material base	S275	S460
Intensity	80 A	230 A	No. of welding processes	1	2
Polarity	DC	AC	Voltage	10 V	30 V
Speed	40 mm/min	240 mm/min	Use of backup	No	Yes
Heat input	0.6 KJ/mm	4.2 KJ/mm	Welding thickness	20 mm	75 mm
Pre-heating	No	150 °C	W. position	PA	PF
Gas flux	10 l/min	25 l/min	Bevel angle	35°	90°

Abbreviations: minimum = Min.; maximum = Max.

All the welded coupons were subjected to extend non-destructive tests, according to EN 17637 [40], EN 17638 [41], and EN ISO 17640 [42]. A total of 14 small indications were found, and consequently the zone was marked and discarded for destructive tests.

2.4. CTOD Test

All 72 CTOD tests were done according to ASTM E1290-08e1c (2008) [12] with standard specimens (single-edge notched bend SE(B) specimen with square B × B cross-section) and the recommended notch [43]. The apparatus used was an oleo-hydraulic dynamic machine (model UFIB-200E-MD5W) configured for a 3-point bending setup and using clip-gauges as the crack growth measuring system.

The testing temperature was in the range of 20–25 °C. As Figure 1 shows, the notch was aligned 1 mm from the fusion line.

The chosen testing method, ASTM E1290-08e1, calculate the CTOD value with the following expression:

$$\delta = \frac{1}{m\sigma_Y}\left[\frac{K^2\left(1-v^2\right)}{E} + \frac{\eta_{CMOD}A^{pl}_{CMOD}}{B(W-a_0)\{1+Z/(0.8a_0+0.2W)\}}\right] \tag{2}$$

where Z is the distance of the front face of the SE(B) specimens to the knife-edge measurement point, A^{pl}_{CMOD} is the plastic area under load from the plastic $CMOD$ curve, and the expression of m is:

$$m = A_0 - A_1\left(\frac{\sigma_{YS}}{\sigma_{ts}}\right) + A_2\left(\frac{\sigma_{YS}}{\sigma_{ts}}\right)^2 - A_3\left(\frac{\sigma_{YS}}{\sigma_{ts}}\right)^3 \tag{3}$$

where

$$A_0 = 3.18 - 0.22\left(\frac{a_0}{W}\right),\ A_1 = 4.32 - 2.23\left(\frac{a_0}{W}\right),\ A_2 = 4.44 - 2.29\left(\frac{a_0}{W}\right),\ A_4 = 2.05 - 1.06\left(\frac{a_0}{W}\right) \tag{4}$$

and

$$\eta_{CMOD} = 3.667 - 2.199\left(\frac{a_0}{W}\right) + 0.437\left(\frac{a_0}{W}\right)^2, \tag{5}$$

Alternatives calculations, formulas, and predictions were studied by [33,44–48].

All the tests were performed in the private laboratory testing facilities of the TAM group (accreditation no. 808/LE1532).

2.5. Results

The data obtained were processed according to the respective test procedures. Finally, for each of the 72 test samples, the results were collected for the explanatory and objective variables. In the Table 3 the results of the testing process are summarized and expressed as the minimum (Min.) and maximum (Max.), giving the range, the average value (Avg.), standard deviation (SD), and coefficient of variation (CV).

Table 3. Experimental phase results summary.

Param.	CTOD [mm]	Mechanical Strength [HV10]	Toughness [J]	Microstructure [μm. Correction]	Chemical Composition [CE]
Min.	0.10	165	76	101	0.22
Max.	2.45	375	278	354	0.53
Avg.	1.24	224.4	183.4	177.0	0.39
SD	0.59	41.87	51.84	55.44	0.05
CV [%]	47.3	18.7	28.3	31.3	14.1

Abbreviations: parameters = Param.; minimum = Min.; maximum = Max.; average value = Avg.; standard deviation = SD; coefficient of variation = CV; crack tip opening displacement = CTOD.

3. Modeling

We observed a set of K variables $X_1, X_2, \ldots, X_K$ in a set of n elements of a population and wanted to summarize the values of the variables and describe their dependency structure. Each of these K variables is called a scalar or univariate variable and the set of these K variables form a vector or multivariate variable. All these values can be represented in a matrix, X, of dimensions $n \times p$, called a data matrix, where each row represents the values of the K variables over the individual i, and each column represents the corresponding scalar variable measured in the n elements of the population. In the element x_{ij}, i denotes the individual and j is the variable.

Next, we proceed to the multivariate analysis of the observations. To do this, we calculate the vector of means $\overline{X} = \begin{bmatrix} \overline{X}_1 & \overline{X}_2 & \cdots & \overline{X}_K \end{bmatrix}$ of dimension p, whose components are the means of each of the p variables and the covariance matrix. From the matrix of centered data $\widetilde{X}$,

$$\widetilde{X} = X - \begin{bmatrix} 1 \\ 1 \\ \vdots \\ 1 \end{bmatrix} \overline{X}, \tag{6}$$

the symmetric and positive semidefinite matrix of covariance $S = \frac{1}{n}\widetilde{X}^T\widetilde{X}$ is calculated.

The objective of describing multivariate data is to understand the dependence between the objective variable and the explanatory variables. For this we studied:

1. The relationship between pairs of variables;
2. Dependence between the objective variable and all the explanatory variables;
3. Dependence between the objective variable and the explanatory ones, but eliminating the effect of some of them.

The pairwise dependence between the variables is measured by the symmetric and positive semidefinite correlation matrix R

$$R = \begin{bmatrix} 1 & r_{12} & \cdots & r_{1K} \\ r_{21} & 1 & \cdots & r_{2K} \\ \vdots & \vdots & \ddots & \vdots \\ r_{K1} & r_{K2} & \cdots & 1 \end{bmatrix}, \quad r_{jk} = \frac{S_{jk}}{S_j S_k} \tag{7}$$

so that there is an exact linear relationship between the variables X_j and X_k if $\left|r_{jk}\right| = 1$.

It may happen that there are variables that are very dependent on others, in which case it is convenient to measure their degree of dependence. Assuming that $Y = X_j$ is the variable of interest, and calling $\hat{Y}$ the variable used to estimate Y, the best linear predictor from the other variables, called the explanatory variables, is:

$$\hat{Y} = \beta_0 + \beta_1 X_1 + \cdots + \beta_K X_K, \tag{8}$$

where the parameter β_i is determined through the data that we have at our disposal. The problem is finding the set of parameters that minimizes $\sum_{i=1}^{n} \left(Y_i - \hat{Y}_i\right)^2$, leading to

$$\begin{aligned} y &= Y - \overline{Y} \\ x_j &= X_j - \overline{X}_j, \quad j = 1, \dots, K \end{aligned} \tag{9}$$

and defining $\hat{y} = \hat{Y} - \overline{Y}$, we have $Y - \hat{Y} = y - \hat{y}$, and Equation (8) can be written as follows

$$\hat{y} = \alpha_0 + \alpha_1 x_1 + \cdots + \alpha_K x_K, \tag{10}$$

Since minimizing $\sum_{i=1}^{n} \left(Y_i - \hat{Y}_i\right)^2$ is equivalent to minimizing $\sum_{i=1}^{n} \left(y_i - \hat{y}_i\right)^2 = \sum_{i=1}^{n} e_i^2$, by deriving this sum with respect to the α_k parameters, we obtain a system of $p - 1$ equations that can be written as follows:

$$\sum_{i=1}^{n} e_i x_{il} \quad l = 1, \dots, K, \quad l \neq j \tag{11}$$

Equation (9) indicates that the prediction errors must not be correlated with the explanatory variables, so that the covariance of both is zero, or else the residual vector must be orthogonal to the space generated by the explanatory variables. By defining the matrix X_R, of size $n \times (p-1)$, obtained by eliminating the column in the matrix $\widetilde{X}$ corresponding to the variable that we want to predict, $y = x_j$, the parameters are calculated by the normal equation system as follows

$$\alpha = \left(X_R^T X_R\right)^{-1} X_R^T y \tag{12}$$

and Equation (10), with these coefficients, is the multiple regression equation between variable $y = x_j$ and the remaining variables x_i, $i \neq j$, $i = 1, \ldots, K$.

To express this result based on the $X_1, \ldots, X_K$ variables of Equation (8), we must consider

$$\beta_i = \alpha_i, \ i = 1, \ldots, K$$
$$\beta_0 = \alpha_0 + \overline{Y} - \sum_{i=1}^{K} \alpha_i \overline{X}_i \tag{13}$$

The square of the multiple correlation coefficient (which can be greater than, less than, or equal to the sum of the squares of the simple correlations between variable y and each of the explanatory variables) [49] between the variable $y = x_j$ and the rest is

$$R_j^2 = 1 - \frac{SS_{resid}}{SS_{total}} = 1 - \frac{1}{s_{jj}s^{jj}}, \tag{14}$$

where $s_{jj} = s_j^2$ is the j-th diagonal element of the covariance matrix S and $s^{jj} = \frac{1}{s_r^2(j)}$ is the j-th diagonal element of the S^{-1} matrix, which represents the residual variance of a regression between the j-th variable and the rest. As each time a variable is added to the model the number of degrees of freedom is reduced and the adjustment is increased, it is necessary to make a correction of this coefficient and calculate the adjusted R_j^2,

$$\overline{R}_j^2 = 1 - \frac{\frac{SS_{resid}}{(n-k)}}{\frac{SS_{total}}{(n-1)}}, \tag{15}$$

where n is the total number of observations and k is the number of model variables; that is, the same calculation is made, but weighted by the degrees of freedom of the residuals, $n - k$, and the model, $n - 1$.

The R-squared $RSQ = \frac{\sum (\hat{y}_i - y_i)^2}{\sum (y_i - \overline{y})^2}$ is a descriptive measure of the predictive capacity of the model, and for a single explanatory variable is the square of the simple correlation coefficient between the two variables.

3.1. Previous Data Processing

Correlation coefficients were determined among the study variables. A high degree of correlation between toughness (CVN) and microstructure was observed, which was strongly supported in the bibliography. This relationship also depends on other variables that have not been considered in this experiment, such as temperature, tension state, or specimen geometry. Therefore, this particular relation between both variables is exclusive to this experiment and cannot be generalized.

Figure 2 shows the correlation and scatterplot diagrams between all the variables (objective and explanatory) taken two-by-two. The kernel density estimation (KDE) representation is also a way to estimate the probability density function of a random variable. A strong correlation can be observed among the CTOD and the explanatory variables, particularly toughness, microstructure, and chemical composition. Excluding the chemical composition, other variables do not seem to follow a normal distribution.

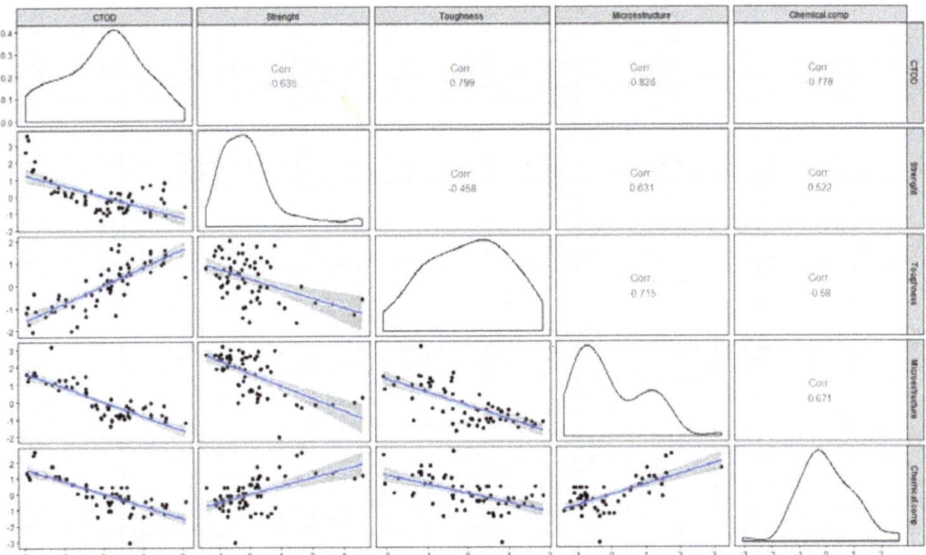

Figure 2. Correlation, kernel density estimation (KDE), and scatterplots (the trendline that best fit linear relation is represented in blue) among the different variables.

Figure 3 shows the quantiles of input samples (explanatory variables) versus standard normal quantiles (theoretical quantiles from a normal distribution). If the distribution of the explanatory variable is normal, the plot will be close to linear. Except for the chemical composition and toughness, the rest of the independent variables (the mechanical strength, called M. Strength onwards, and microstructure) do not seem to follow a normal distribution, so it would be advisable to make a transformation (for example, logarithmic type) before carrying out a multiple regression analysis. This can be explained by the observation of the KDE of the corresponding variable in Figure 2, where the M. Strength variable shows a positive skewness towards lower values and the microstructure shows a slightly bimodal distribution (this effect is eliminated through a logarithmic transformation after the outlier exclusion).

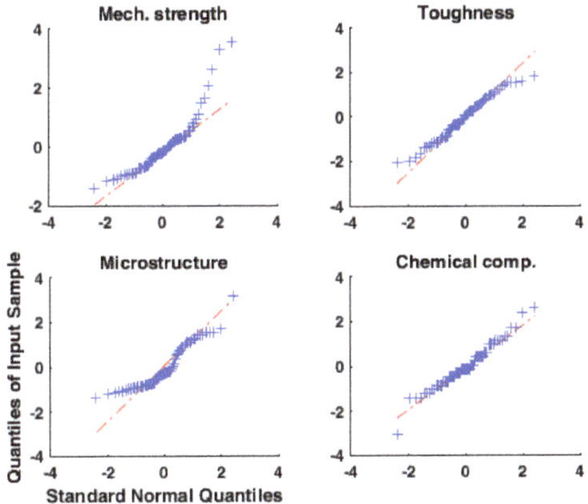

Figure 3. Quantiles of input sample versus standard normal quantiles.

With the aim of discarding the outliers that could influence observations, the Mahalanobis distance was used [49,50] for their detection and ten complete data sets were excluded (14%).

3.2. Linear Regression Models

3.2.1. Linear Model 1

Here, Y is considered as the study variable that may be linearly related with K explanatory variables $X_1, X_2, \cdots, X_K$ through $\beta_0, \beta_1, \beta_2, \cdots, \beta_K$ (regression coefficients). A multiple linear regression model can be written as:

$$Y = \beta_0 + \beta_1 X_1 + \beta_2 X_2 + \cdots + \beta_K X_K + e \tag{16}$$

where e is the difference between the fitted relationship and the observations [51].

Using Equations (12) and (13), the values of the parameters are calculated. In Table 4, the coefficients for the multiple linear regression (Equation (16)) can be found. It can be seen that all coefficients are significantly different from zero, but toughness is the variable with the highest absolute value. In this case, the number of observations is 63, and the error degrees of freedom is 58.

Table 4. Multiple linear model 1 $Y = \beta_0 + \beta_1 X_1 + \beta_2 X_2 + \beta_3 X_3 + \beta_4 X_4$.

Parameters	Estimate	SE	p-Value
β_0	1.2202	0.022731	1.0629×10^{-46}
$\beta_1 - M.\ Str.$	-0.080323	0.035816	2.8756×10^{-2}
$\beta_2 - Tough.$	0.22424	0.040094	6.3259×10^{-7}
$\beta_3 - Micros.$	-0.12972	0.047887	8.8595×10^{-3}
$\beta_4 - C.\ Comp.$	-0.19243	0.038258	5.0415×10^{-6}
F-Statistic p-value	-	-	1.4×10^{-24}

The root mean square error (RMSE) is 0.216, which when compared to the range of the values of Y results in:

$$\frac{RMSE}{(Y_{MAX} - Y_{MIN})} = 0.1048 \approx 10\%, \tag{17}$$

Which provides an estimate of the possible error obtained from the real values of the CTOD variable. In Figure 2, it can be observed that the correlation coefficient between CTOD and toughness is 0.799. Considering all the independent variables the R-squared (RSQ) is 0.866, and the adjusted RSQ value is 0.856, so there is a limited improvement from considering the CTOD toughness (or CTOD microstructure) correlation.

Henceforth, for the models shown, the t-statistic (tStat) and F-statistic will be calculated and included. The first of them, tStat, calculated as estimated or standard error (SE), tests the null hypothesis that the corresponding coefficient is zero against the alternative that it is different from zero. To evaluate this coefficient, the corresponding p-value associated with a Student's t distribution (for n observations) is calculated and compared with a confidence interval of 95%. If the p-value is less than 0.05, we can conclude that the variable is significant for the model.

Analogously, the F-statistic, calculated as:

$$F = \frac{\sum_{i=1}^{n} \frac{(\hat{y}_i - \overline{y})^2}{(p-1)}}{\sum_{i=1}^{n} \frac{(y_i - \hat{y}_i)^2}{(n-p)}} \tag{18}$$

tests the null hypothesis that one or more of the regression coefficients are significantly different from zero (meaning a significant linear regression relationship exists for the whole model). This value is compared with an F-distribution for a given confidence interval (95%) and is evaluated in the same way as the t-statistic (associated p-value less than 0.05). The F-distribution is more appropriate than Chi-square tests for small data sets [52].

Two different methods were used to verify that the obtained model was independent of the chosen data population: cross-validation and training-test samples.

The cross-validation was calculated with the *LeaveMout* method (see *crossvalind* Matlab function) with an M value of 1, which randomly selects one value and excludes it from the evaluation. This process is repeated 50 times and helps to verify that the statistical analysis is independent of the data set. The number of observations was 62, with $RMSE = 0.218$, $RSQ = 0.866$, and adjusted $RSQ = 0.856$. The results are shown in Table 5.

Table 5. Cross-validation results.

Parameters	Estimate	SE	p-Value
β_0	1.2215	0.027668	8.971×10^{-46}
$\beta_1 - M.\ Str.$	-0.08113	0.036144	2.8692×10^{-2}
$\beta_2 - Tough.$	0.22244	0.040667	1.0436×10^{-6}
$\beta_3 - Micros.$	-0.13334	0.049165	8.8225×10^{-3}
$\beta_4 - C.\ Comp.$	-0.19006	0.039041	9.3064×10^{-6}
F-Statistic p-value	-	-	3.5×10^{-24}

The training test was done considering a set of 500 executions of samples from 50 observations (randomly selected from the whole data set) and test samples from 13 data sets. The averages of all RMSE and RSQ results are $\overline{RMSE} = 0.2275$ and $\overline{RSQ} = 0.8284$, respectively.

Table 6 contains the values of RSQ and RMSE obtained with the reference model (linear model 1), cross-validation, and training test. As the values are similar (less than 5% discrepancy), we can conclude that the relation between the CTOD and the explanatory variables is independent of the data set.

Table 6. Comparison of the values of R-squared (RSQ) and root mean square error (RMSE).

Parameters	RSQ	RMSE
Linear Model 1	0.866	0.216
Cross reference	0.866	0.218
Training-Test	0.828	0.227

3.2.2. Linear Model 2

The significance of all variables was checked for all the explanatory variables, but it was observed that the microstructure was highly correlated with toughness. For that reason, a new model (linear model 2) was proposed, where the microstructure was eliminated from the original model.

$$Y = \beta_0 + \beta_1 X_1 + \beta_2 X_2 + \beta_4 X_4 \tag{19}$$

Table 7 shows the values of the parameters calculated for linear model 2, and the adjustment obtained ($RMSE = 0.227$, $RSQ = 0.849$, and adjusted $RSQ = 0.841$) was similar to the previous one (linear model 1).

Table 7. Multiple linear regression model 2 $Y = \beta_0 + \beta_1 X_1 + \beta_2 X_2 + \beta_4 X_4$.

Parameters	Estimate	SE	p-Value
β_0	1.2202	0.028657	5.353×10^{-46}
$\beta_1 - M.\ Str.$	-0.11829	0.034686	1.1768×10^{-3}
$\beta_2 - Tough.$	0.27944	0.036337	1.8301×10^{-10}
$\beta_4 - C.\ Comp.$	-0.22775	0.03785	1.2105×10^{-7}
F-Statistic p-value	-	-	3.79×10^{-24}

3.2.3. Linear Models 3 and 4

As the value of parameter β_1 (coefficient of the mechanical strength) in linear model 1 was small compared to the values of the rest of the parameters, it was that the corresponding variable be eliminated to obtain a new model (linear model 3), considering that its contribution to the value of the CTOD variable was small. The values of the coefficients of linear model 3 are represented in Table 8.

Table 8. Linear regression model 3 $Y = \beta_0 + \beta_2 X_2 + \beta_3 X_3 + \beta_4 X_4$.

Parameters	Estimate	SE	p-Value
β_0	1.2202	0.028146	1.9133×10^{-46}
$\beta_2 - Tough.$	0.22225	0.04143	1.4246×10^{-6}
$\beta_3 - Micros.$	-0.17174	0.045549	3.793×10^{-4}
$\beta_4 - C.\ Comp.$	-0.20715	0.038957	1.961×10^{-6}
F-Statistic p-value	-	-	1.32×10^{-24}

The quality of the adjustment is almost similar to that of the model with the four independent variables, with $RMSE = 0.223$ and $RSQ = 0.854$.

Figure 4 shows the residuals of linear model 3, which can be considered as normally distributed.

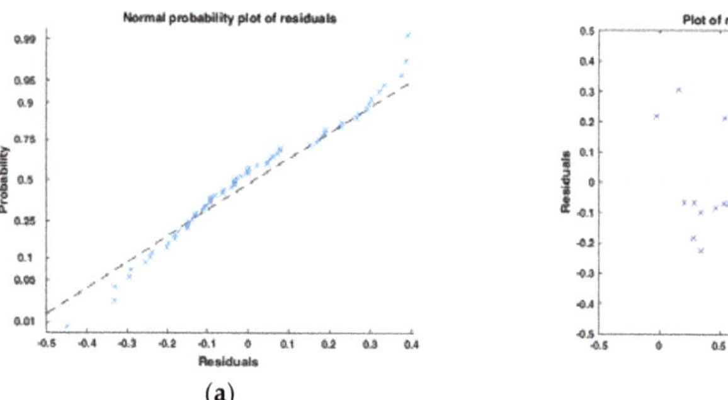

(a)   (b)

Figure 4. Normal distribution of the residuals without variables. (**a**) Normal probability plot of residuals (**b**) Plot of residuals vs. fitted values.

Finally, a new model (linear model 4) is adopted considering the square of the first variable ($M.\ Str^2.$), and the contribution of the independent variables to the variable CTOD is checked (see Table 9). In this case, the coefficient of determination $RSQ = 0.874$ is larger than in the purely linear model.

Table 9. Linear regression model 4 $Y = \beta_0 + \beta_1 X_1^2 + \beta_2 X_2 + \beta_3 X_3 + \beta_4 X_4$.

Parameters	Estimate	SE	p-Value
β_0	1.2584	0.029113	8.0522×10^{-46}
$\beta_1 - M.\ Str^2.$	-0.038872	0.012639	3.2021×10^{-3}
$\beta_2 - Tough.$	0.21641	0.038792	6.6704×10^{-7}
$\beta_3 - Micros.$	-0.1562	0.042897	5.7976×10^{-4}
$\beta_4 - C.\ Comp.$	-0.19142	0.036789	2.68×10^{-6}
F-Statistic p-value	-	-	1.97×10^{-25}

Other tests have been done with different interactions between variables, but they do not improve the results.

3.3. Multivariate Adaptative Regression Splines (MARS)

Multivariate adaptive regression splines (MARS) is a non-parametric modeling method that extends the linear model (incorporating nonlinearities and interactions). It is a generalization of the recursive partitioning regression (RPR), which splits up the space of the explanatory variables into different subregions. MARS generates cut points for the variables. These knots are identified through baseline functions, which indicates the beginning and end of a region.

In each region in which the space is divided, a base linear function of one variable is adjusted. The final model is constituted from a combination of the generated base functions [53].

The general expression of the model is:

$$\hat{Y} = \sum_{i=1}^{k} c_i B_i(x) , \tag{20}$$

where c_i is the constant coefficient and B_i is the base function.

A MARS model was applied using cubic splines. This method considers nonlinear relationships among the CTOD variable and the explanatory ones using a spline adjustment, obtaining a $RSQ = 0.86$ and $RMSE = 0.16$. With a training sample of 50 data sets and test sample of 13, the results were $RSQ = 0.84$ and $RMSE = 0.26$. Additional information may be found in Figure 5, where the MARS model is plotted for two of the explanatory variables and two anaylsis of variance (ANOVA) functions (this visualizes the contribution of the ANOVA functions for the pairs CTOD-M. Strength and CTOD-microstructure in the MARS model).

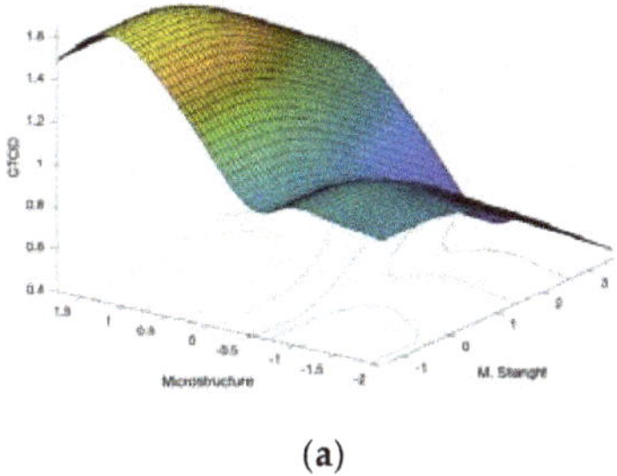

(**a**)

Figure 5. *Cont.*

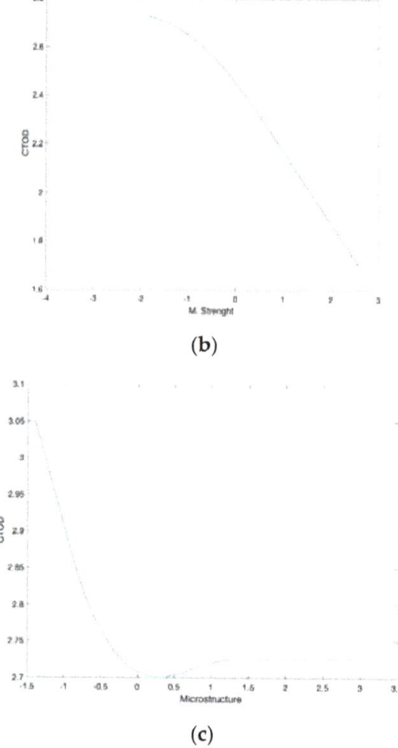

(b)

(c)

Figure 5. (**a**) Multivariate adaptive regression splines (MARS) model plot for two of the explanatory variables together with its knot locations (up) and (**b**) the analysis of variance (ANOVA) function for the pairs CTOD-M. Strength (left) and (**c**) CTOD-microstructure (right) (using ARESLab toolbox: Jekabsons G., ARESLab: Adaptive Regression Splines Toolbox for Matlab/Octave, 2016, available at http://www.cs.rtu.lv/jekabsons/).

Again, these values do not improve on those obtained with previous models.

3.4. Other Models

Other models were studied in order to observe a possible improvement with respect to the initial model (linear model 1).

In the first place, we proposed a generalized linear model considering a Gaussian distribution and an identity linking function, the parameters for which are included in Table 10 (Generalized linear regression model 1—GLM1). It is noted that the p-value of the mechanical strength is greater than 0.05, therefore, the variable X_1 (mechanical strength) may not be significant.

Table 10. Generalized linear regression model 1 (GLM1) $Y = \beta_0 + \beta_1 X_1 + \beta_2 X_2 + \beta_3 X_3 + \beta_4 X_4$.

Parameters	Estimate	SE	p-Value
β_0	1.2202	0.027231	1.0629×10^{-46}
$\beta_1 - M.\ Str.$	-0.08032	0.035816	2.8756×10^{-2}
$\beta_2 - Tough.$	0.22424	0.040094	6.3259×10^{-7}
$\beta_3 - Micros.$	-0.12972	0.047887	8.859×10^{-3}
$\beta_4 - C.\ Comp.$	-0.19243	0.038259	5.0415×10^{-6}
F-Statistic p-value	-	-	1.4×10^{-24}

For this reason, a generalized linear model was calculated without the mechanical strength influence (GLM2), whose results are shown in Table 11, with $RSQ = 0.675$ and $RMSE = 0.3396$ obtained. These values do not improve on those obtained with previous models.

Table 11. Generalized linear regression model 2 (GLM2) $Y = \beta_0 + \beta_2 X_2 + \beta_3 X_3 + \beta_4 X_4$.

Parameters	Estimate	SE	p-Value
β_0	1.0169	0.043522	3.701×10^{-27}
$\beta_2 - Tough.$	-0.032814	0.037709	3.8872×10^{-2}
$\beta_3 - Micros.$	0.3569	0.059269	2.6872×10^{-7}
$\beta_4 - C. Comp.$	0.15866	0.031816	9.1927×10^{-6}
F-Statistic p-value	-	-	3.93×10^{-13}

In the second place, we considered a regression tree model [54]. To make a prediction for a given observation, we used the mean (or the mode) of the observations that were in the same region of the multidimensional space of predictors. The rules that were used to divide the predictor space can be represented as a tree [55].

The order of importance of the predictive variables, from highest to lowest, is microstructure, toughness, mechanical strength, and chemical composition. Therefore, the variable microstructure is the one that provides the value that maximizes the information about the dependent variable (CTOD) if it is smaller than 0.26, otherwise it is the toughness that carries more information. Nevertheless, the values associated with each subtree for the training sample (13) are between 3 to 7 times bigger than those of the test sample (50), which indicates bad behavior of the model.

4. Results and Discussion

After having compared the previous model, due to the simplicity and reasonable accuracy, and despite the unbalanced weigh of the different variable's parameters (β_i), linear model 1 ($RSQ = 0.866$) is proposed as a predictive model of the values of CTOD.

The standardized model can be expressed as:

$$CTOD = 1.2202 \quad -0.080323[M.Str] + 0.22424[Tough] - 0.12972[Micros] \\ -0.19243[C.Comp] \tag{21}$$

where [.] represents the standardized values of the corresponding variable. The standardization process should be reverted to allow the use of the testing data directly:

$$\frac{(CTOD[mm]-1.0269)}{0.586} = 1.2202 - 0.080323\frac{(M.Str[HV10]-221.09)}{41.87} \\ +0.22424\frac{(Tough[J]-175.421)}{51.84} - 0.12972\frac{(Micros[\mu m]-169.23)}{55.44} \\ -0.19243\frac{(CE-0.3831)}{0.054} \tag{22}$$

where $CTOD[mm]$ is the value of the crack tip opening displacement, expressed in mm. $M.Str[HV10]$ is the effect of the mechanical strength of the material as the average of three hardness measurements expressed in $[HV10]$ units. $Tough[J]$ is the average value of the two Charpy V-notch measurements, with subsized specimens extracted from the interest zone and expressed in Joules. $Micros[\mu m]$ is the average size of the metallic grains expressed in μm. Finally, CE is the effect of the chemical composition as the carbon equivalent calculated with the equation proposed in AWS D1.1 (adimensional).

5. Conclusions

The use of multivariate analysis has been proven viable for relating complex fracture mechanics parameters to well-known material properties. The industrial suitability of the methodology depends on the experimental set, specifically the availability of samples, the number of tests, and the choice of variables.

These chosen variables are significantly related with the CTOD (see p-value for linear regression model 1). Also, there is well-known experience within the manufacturing industry relating these variables with actual changes during the welding process. As an example, there is a wide background of knowledge on how the shielding gas, the welding speed, or the bead scheme affect the grain size or the hardness of a given welded joint. Using the proposed model, it is possible for the industry to transfer this knowledge on how these variables may affect the CTOD value.

The final model is precise and functional, with an estimated error of ~10% (within the limits covered by the experimental set). This error is compatible with the current uncertainty of the CTOD testing process. Besides, the model is not dependent on which subgroup of data is used for the modeling process. It is proposed to use this final model predictively, using the results of the tests for the explanatory variables (it is cheaper, simpler, and more available than the CTOD) to compute the CTOD value estimator. If this value (considering the mentioned error) is greater than the critical value (acceptance criteria) specified in the design code, rule, or standard, the expensive CTOD test can be dispensed.

The usefulness of the model has been proven within the limits of the experimental set for offshore steel welded joints of high thickness. Nevertheless, the influences of other variables not explicitly considered in this work were not tested, even for the mentioned category, and are out of the scope of the presented model. Future developments of the model could include, among others, the influence of testing temperature, different positioning or shape of the notch, post-weld heat treatments, or type of failure category (brittle, ductile-brittle, and ductile).

Author Contributions: Conceptualization, Á.P.V. and A.B.S.; methodology, Á.P.V. and A.B.S.; software, Á.P.V. and Z.F.M.; validation, Z.F.M. and M.M.F.; formal analysis, Á.P.V.; investigation, Á.P.V. and A.B.S.; resources, Á.P.V. and A.B.S.; data curation, Z.F.M.; writing—original draft preparation, Á.P.V.; writing—review and editing, Á.P.V., Z.F.M., and M.M.F.; visualization, Á.P.V.; supervision, A.B.S.

Funding: This research did not receive any specific grant from funding agencies in the public, commercial, or not-for-profit sectors.

Acknowledgments: The authors would like to thank the private laboratory TAM for access to testing equipment, software, and mediation for the transfer of testing samples.

Conflicts of Interest: The authors declare no conflict of interest.

References

1. Rodrigues, P.E.; Wong, W.K.; Rogerson, J.H. Weld Defect Distributions in Offshore Platforms and Their Relevance to Reliability Studies, Quality Control and In-Service Inspection. In Proceedings of the Offshore Technology Conference, Houston, TX, USA, 5–8 May 1980.
2. American Bureau of Shipping. *Design Standards for Offshore Wind Farms*; American Bureau of Shipping: Huston Texas, USA, 2011.
3. Smith, M. *Fitness for Service*, API 579-1/ASME FFS-1. 2007.
4. Offshore Standard Det Norske Veritas DNV-OS-C101. Available online: https://www.scribd.com/document/50782302/DNV-OS-C101-Design-of-offshore-steel-structures-general-LRFD-Method-October-2008 (accessed on 27 September 2019).
5. Zhu, X.; Joyce, A. Review of fracture toughness (G, K, J, CTOD, CTOA) testing and standardization. *Eng. Fract. Mech.* **2012**, *85*, 1–46.
6. Irwin, G.R.; Kies, J.A. Critical energy rate analysis for fracture strength. *Weld. J. Res. Suppl.* **1954**, *19*, 193–198.
7. Wells, A.A. Application of fracture mechanics at and beyond general yielding. *Br. Weld. J.* **1963**, *10*, 563–570.
8. Rice, J.R. A path independent integral and the approximate analysis of strain concentration by notches and cracks. *J. Appl. Mech.* **1968**, *35*, 379–386.
9. Newman, J.C.; James, M.A.; Zerbst, U. A review of the CTOA/CTOD fracture criterion. *Eng. Fract. Mech.* **2003**, *70*, 371–385.

10. Akourri, O.; Louah, M.; Kifani, A.; Gilgert, G.; Pluvinage, G. The effect of notch radius on fracture toughness JIc. *Eng. Fract. Mech.* **2000**, *65*, 491–505.

11. Li, X.; Song, Y.; Ding, Z.; Bao, S.; Gao, Z. A modified correlation between KJIC and Charpy V-notch impact energy of Chinese SA508-III steel at the upper shelf. *J. Nucl. Mater.* **2018**, *505*, 22–29.

12. ASTM E1290-08e1c. *Standard Test Method for Crack tip Opening Displacement (CTOD) Fracture Toughness Measurement*; ASTM International: West Conshohocken, PA, USA, 2008.

13. Rencher, A.C. *Methods of Multivariate Analysis*; John Wiley & Sons: Hoboken, NJ, USA, 2002.

14. Dunne, D.; Tsuei, H.; Sterjovski, Z. Artificial neural networks for modeling of the impact toughness of steel. *ISIJ Int.* **2004**, *44*, 1599–1607.

15. Haque, M.E.; Sudhakar, K.V. ANN back-propagation prediction model for fracture toughness in microalloy steel. *Int. J. Fatigue* **2002**, *24*, 1003–1010.

16. ASM International. *ASM Handbook*; ASM International: Materials Park, OH, USA, 1996; Volume 19.

17. Castelluccio, G.M.; McDowell, D.L. Microstructure-sensitive small fatigue crack growth assessment: Effect of strain ratio, multiaxial strain state, and geometric discontinuities. *Int. J. Fatigue* **2016**, *82*, 521–529.

18. Chen, Y.Q.; Pan, S.P.; Zhou, M.Z.; Yi, D.Q.; Xu, D.Z.; Xu, Y.F. Effects of inclusions, grain boundaries and grain orientations on the fatigue crack initiation and propagation behavior of 2524-T3 Al alloy. *Mater. Sci. Eng.* **2013**, *580*, 150–158. [CrossRef]

19. Cheng, T.C.; Yu, C.; Yang, T.C.; Huang, C.Y.; Lin, H.C.; Shiue, R.K. Microstructure and Impact Toughness of Offshore Steel. *Arch. Metall. Mater.* **2018**, *63*, 167–172.

20. Pegues, J.W.; Roach, M.D.; Shamsaei, N. Influence of microstructure on fatigue crack nucleation and microstructurally short crack growth of an austenitic stainless steel. *Mater. Sci. Eng.* **2017**, *707*, 657–667. [CrossRef]

21. Wei, L.; Pan, Q.; Huang, H.; Feng, L.; Wang, Y. Influence of grain structure and crystallographic orientation on fatigue crack propagation behavior of 7050 alloy thick plate. *Int. J. Fatigue* **2014**, *66*, 55–64. [CrossRef]

22. Yuan, H.; Zhang, W.; Castelluccio, G.M.; Kim, J.; Liu, Y. Microstructure-sensitive estimation of small fatigue crack growth in bridge steel welds. *Int. J. Fatigue* **2018**, *112*, 183–197. [CrossRef]

23. Yuan, H.; Zhang, W.; Kim, J.; Liu, Y. A nonlinear grain-based fatigue damage model for civil infrastructure under variable amplitude loads. *Int. J. Fatigue* **2017**, *104*, 389–396. [CrossRef]

24. ASTM E112. *Standard Test Methods for Determining Average Grain Size*; ASTM International: West Conshohocken, PA, USA, 2013.

25. ASTM E3—11. *Standard Guide for Preparation of Metallographic Specimens*; ASTM International: West Conshohocken, PA, USA, 2018.

26. ASTM E407—07. *Standard Practice for Micro-Etching Metals and Alloys*; ASTM International: West Conshohocken, PA, USA, 2015.

27. Talaş, Ş. The assessment of carbon equivalent formulas in predicting the properties of steel weld metals. *Mater. Des.* **2010**, *31*, 2649–2653. [CrossRef]

28. Lee, S.G.; Lee, D.H.; Sohn, S.S.; Kim, W.G.; Um, K.K.; Kim, K.S.; Lee, S. Effects of Ni and Mn addition on critical crack tip opening displacement (CTOD) of weld-simulated heat-affected zones of three high-strength low-alloy (HSLA) steels. *Mater. Sci. Eng.* **2017**, *697*, 55–65. [CrossRef]

29. Yurioka, N. Physical Metallurgy of Steel Weldability. *ISIJ Int.* **2001**, *41*, 566–570. [CrossRef]

30. American Welding Society D1.1/D1.1M. *Structural Welding Code Steel*; American Welding Society: Miami, FL, USA, 2010.

31. Anderson, T.L. *Fracture Mechanics: Fundamentals and Applications*, 3rd ed.; Taylor & Francis: Boca Raton, FL, USA, 2005.

32. ASTM E92. *Standard Test Methods for Vickers Hardness and Knoop Hardness of Metallic Materials*; ASTM International: West Conshohocken, PA, USA, 2017.

33. Khor, W.; Moore, P.L.; Pisarski, H.G.; Haslett, M.; Brown, C.J. Measurement and prediction of CTOD in austenitic stainless steel. *Fatigue Fract. Eng. Mater. Struct.* **2016**, *39*, 1433–1442. [CrossRef]

34. Maropoulos, S.; Ridley, N.; Kechagias, J.; Karagiannis, S. Fracture toughness evaluation of a H.S.L.A. steel. *Eng. Fract. Mech.* **2004**, *71*, 1695–1704. [CrossRef]

35. ASME BPVC IX. *Boiler Pressure Vessel Code (BPVC) Welding, Brazing, and Fusing Qualifications*; The American Society of Mechanical Engineers (ASME): New York, NY, USA, 2017.

36. ASTM E23. *Standard Test Methods for Notched Bar Impact Testing of Metallic Materials*; ASTM International: West Conshohocken, PA, USA, 2018.

37. Yang, Y.Y.; Mahfouf, M.; Panoutsos, G. Probabilistic Characterization of Model Error Using Gaussian Mixture Model—with Application to Charpy Impact Energy Prediction for Alloy Steel. *Control Eng. Pract.* **2012**, *20*, 82–92. [CrossRef]

38. EN 10225. *Weldable Structural Steels for Fixed Offshore Structures*; European Committee for Standardization: Brussels, Belgium, 2009.

39. EN ISO 10025. *Hot Rolled Products of Structural Steels*; European Committee for Standardization: Brussels, Belgium, 2006.

40. EN ISO 17637. *Non-Destructive Testing of Welds—Visual Testing of Fusion-Welded Joints*; European Committee for Standardization: Brussels, Belgium, 2017.

41. EN ISO 17638. *Non-Destructive Testing of Welds—Magnetic Particle Testing*; European Committee for Standardization: Brussels, Belgium, 2017.

42. EN ISO 17640. *Non-Destructive Testing of Welds—Ultrasonic Testing—Techniques, Testing Levels, and Assessment*; European Committee for Standardization: Brussels, Belgium, 2011.

43. Ávila, J.A.; Lima, V.; Ruchert, C.O.; Mei, P.R.; Ramírez, A.J. Guide for Recommended Practices to Perform Crack Tip Opening Displacement Tests in High Strength Low Alloy Steels. *Soldag. Inspeção* **2016**, *21*, 290–302. [CrossRef]

44. Antunes, F.V.; Branco, R.; Prates, P.A.; Borrego, L. Fatigue crack growth modeling based on CTOD for the 7050-T6 alloy. *Fatigue Fract. Eng. Mater. Struct.* **2017**, *40*, 11. [CrossRef]

45. Antunes, F.V.; Rodrigues, S.M.; Branco, R.; Camas, D. A numerical analysis of CTOD in constant amplitude fatigue crack growth. *Theor. Appl. Fract. Mech.* **2016**, *85*, 45–55. [CrossRef]

46. Janssen, M.; Zuidema, J.; Wanhill, R.J.H. *Fracture Mechanics*, 2nd ed.; Spon Press: New York, NY, USA, 2004.

47. Kawabata, T.; Tagawa, T.; Sakimoto, T.; Kayamori, Y.; Ohata, M.; Yamashita, Y.; Tamura, E.I.; Yoshinari, H.; Aihara, S.; Minami, F.; et al. Proposal for a new CTOD calculation formula. *Eng. Fract. Mech.* **2016**, *159*, 16–34. [CrossRef]

48. Khor, W.L.; Moore, P.; Pisarski, H.; Brown, C. Comparison of methods to determine CTOD for SENB specimens in different strain hardening steels. *Fatigue Fract. Eng. Mater. Struct.* **2017**, *41*, 551–564. [CrossRef]

49. Cuadras, C.M. *Métodos de Análisis Multivariante*; Eunibar: Barça, Barcelona, 1981.

50. Everitt, B.S. *Cluster Analysis*; Edward Arnold: London, UK, 1993.

51. Rao, C.R.; Toutenburg, H.; Shalabh Heumann, C. *Linear Models and Generalizations*; Springer Series in Statistics; Springer: Berlin/Heidelberg, Germany, 2008.

52. Goldstein, H. *Introduction to F-testing in Linear Regression Models*; Lecture note of the Department of Statistics, University of Oslo: Oslo, Norway, 2014.

53. Vanegas, J.; Vásquez, F. Multivariate Adaptative Regression Splines (MARS), una alternativa para el análisis de series de tiempo. *Gac. Sanit.* **2017**, *31*, 235–237. [CrossRef]

54. Berk, R.A. *Classification and Regression Trees (CART). Statistical Learning from a Regression Perspective*; Springer Series in Statistics; Springer: New York, NY, USA, 2008.

55. Seoane, J.; Carmona, C.P.; Tarjuelo, R.; Planillo, A. *Árboles de Regresión y Clasificación*; Análisis Bioestadístico con Modelos de Regresión en R, UAM: Mexico City, Mexico, 2014.

Article

Understanding and Modeling Climate Impacts on Photosynthetic Dynamics with FLUXNET Data and Neural Networks [†]

Nanyan Zhu [1], Chen Liu [2], Andrew F. Laine [3] and Jia Guo [4,*]

[1]   Biological Sciences, Columbia University, New York City, NY 10027, USA; nz2305@columbia.edu
[2]   Electrical Engineering, Columbia University, New York City, NY 10027, USA; cl3760@columbia.edu
[3]   Biomedical Engineering, and Radiology, Columbia University, New York City, NY 10027, USA; al418@columbia.edu
[4]   Psychiatry, and Mortimer B. Zuckerman Mind Brain Behavior Institute, Columbia University, New York City, NY 10027, USA
*   Correspondence: jg3400@columbia.edu
†   The code is publicly available on https://github.com/RosalieZhu/DL-CO2. This paper is adapted from a conference version presented on the 2019 9th International Conference on Future Environment and Energy.

Received: 1 February 2020; Accepted: 6 March 2020; Published: 12 March 2020

Abstract: Global warming, which largely results from excessive carbon emission, has become an increasingly heated international issue due to its ever-detereorating trend and the profound consequences. Plants sequester a large amount of atmospheric CO_2 via photosynthesis, thus greatly mediating global warming. In this study, we aim to model the temporal dynamics of photosynthesis for two different vegetation types to further understand the controlling factors of photosynthesis machinery. We experimented with a feedforward neural network that does not utilize past histories, as well as two networks that integrate past and present information, long short-term memory and transformer. Our results showed that one single climate driver, shortwave radiation, carries the most information with respect to prediction of upcoming photosynthetic activities. We also demonstrated that photosynthesis and its interactions with climate drivers, such as temperature, precipitation, radiation, and vapor pressure deficit, has an internal system memory of about two weeks. Thus, the predictive model could be best trained with historical data over the past two weeks and could best predict temporal evolution of photosynthesis two weeks into the future.

Keywords: neural network; deep learning; climate; photosynthesis; ecology; FLUXNET

---

1. Introduction

Climate change, specifically global warming, has long been considered a pressing issue to the international society as it disrupts the stability of the ecosystem and threatens the prosperity of mankind [1]. Anthropogenic activities that yield excessive carbon emission, such as fossil fuel burning, industrialization and animal husbandry, have been adversely affecting the ecosystem by raising the global temperature. The dynamics of greenhouse gases, such as CO_2, play pivotal roles in controlling the radiative forcing and the energy balance of the whole earth system [2]. Therefore, understanding the full cycle, especially the sources and sinks, of atmospheric CO_2 is critical to better control the CO_2 concentrations in the future to a reasonable extent.

Photosynthesis pathway is the largest land surface CO_2 sink that sequester atmospheric CO_2 into vegetation biomass and stores them as living biomass. The magnitude of photosynthetic carbon sequestration pathway is around 120–130 Pg C per year (1 Pg = 10^{15} grams) [3]. Ecosystem respiration, including autotrophic and heterotrophic, consumes most of the photosynthetic carbon sink, cycles it back into the atmosphere, and results in a much smaller net carbon sink compared with gross carbon input from photosynthesis. The balance between photosynthesis carbon input and ecosystem reparation carbon output determines the fate of atmospheric CO_2 concentration and mitigates the anthropogenic CO_2 emissions, such as from fossil fuel and biomass burning [4].

Given the significant role of photosynthesis carbon uptake in determining the global carbon cycle, atmospheric CO_2 concentrations, radiative balance of earth system, and global warming, it is crucial to gain mechanistic understanding of photosynthesis machinery and its relationship to climate factors such as temperature, precipitation, and radiation. Observational networks, such as FLUXNET [5], have been established globally to measure and understand various different components of land surface carbon cycle including photosynthesis. At leaf scale, photosynthesis reaction is carried out by Ribulose-1,5-bisphosphate carboxylase oxygenase (RuBisCO enzyme), which combines CO_2 and water molecules to generate carbohydrate products. It occurs at two stages: (1) first, light-dependent reactions capture the energy of light and store in adenosine triphosphate (ATP) and nicotinamide adenine dinucleotide phosphate hydrogen (NADPH); (2) second, light-independent reactions capture and reduce carbon dioxide. This biological reaction relies on substrate concentration (CO_2 and water), depends on the activity of temperature-sensitive RUBISCO enzyme, and is driven by solar energy. Thus, theoretically, one is able to build an effective photosynthesis model that takes all those important factors into account and predicts the magnitude of land surface photosynthesis given relevant climate drivers.

Historically, mechanistic models have been used to study the dynamics of land surface photosynthesis and its relevance to the fate of global warming. These models are based on either Monteith law of light use efficiency [6] or Farquhar photosynthesis modeling framework [7]. Alternatively, the photosynthesis rate could also be estimated by data-driven machine learning models that are trained on a large amount of observed photosynthesis data.

The latter approach is gaining popularity because of the growth of Eddy covariance observational networks that collect photosynthetic data on a daily basis over the years, along with the advancements of effective machine learning techniques. Data-driven models have been applied to a wide-range of areas recently, and have demonstrated their robustness in multiple tasks. Methods such as feedforward neural network (FFNN) [8], random forest [9], model trees ensemble [10–12] and support vector regression [13,14] have been utilized to estimate land surface–atmosphere fluxes from site level to regional or global scales [3,11,12,14–19]. Adaptive neuro-fuzzy inference system and general regression neural network have also been used to estimate daily carbon fluxes in forest ecosystems [20]. However, very little research has yet leveraged algorithms that specialize in the integration of temporal dependencies, such as hidden Markov model (HMM), long-short term memory (LSTM) [21] or transformer [22]. Thus, in this study, we employed LSTMs and transformers to predict photosynthetic activities from the past history of climate drivers, and compared their performances against an FFNN counterpart. Our intuition is that the temporal-aware neural network models shall be able to capture the multi-day dynamics of photosynthesis, thus yielding better predictions.

Gross primary product (GPP), a measure of the amount of energy that plants can trap from the sun, is used in this study as a correlate to the rate of photosynthesis. The objective of this study is to model the temporal dynamics of plant photosynthesis in terms of GPP with three different neural network architectures. With appropriately designed model experiments, we also aim to explore and understand the impact of different climate drivers on photosynthesis, as well as the natural ecosystem memory of the photosynthetic activities.

2. Results

We first modeled the dynamics of GPP across four different sites using Feedforward Neural Networks (FFNNs), one for each site and trained independently. The model architecture, formulas, and other details will be covered in the Materials and Methods section. In principle, each prediction of GPP on any day is made by the model based on the six climate drivers including GPP from the previous day. Afterwards, a leave-one-out experiment was conducted for climate drivers, where one driver is excluded from the input at a time, in the hope to draw inferences on which factors are the most influential on predicting GPP. The results are shown in the top panels in Figure 1. The experimental settings for subsequent models were the same unless otherwise specified.

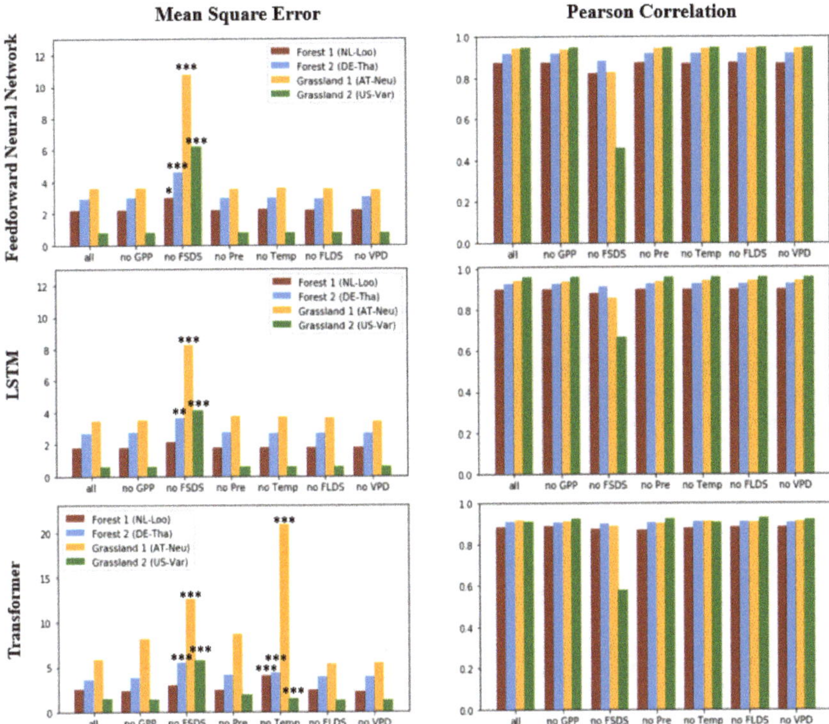

Figure 1. The gross primary product (GPP) prediction performance of (1) the feedforward neural network (FFNN), where only the current climate drivers are utilized; (2) the long short-term memory (LSTM), where current and past climate drivers are taken into account; and (3) the transformer, where current and past climate drivers are comprehensively encoded. A leave-one-out experiment is also conducted for each model, where one of the climate drivers are excluded from the input. Within each neural network candidate, a one-way analysis of variance (ANOVA) followed by a post-hoc Tukey's honestly significant difference (HSD) test [23] is performed to identify the climate drivers without which the model performance would deteriorate. Asteroids indicate the ones whose square errors significantly deviate from those of the version with all climate drivers provided. One, two and three asteroids respectively corresponds to HSD p-value of less than 0.05, 0.01, and 0.001. GPP: gross primary product; FSDS: shortwave radiation; Pre: precipitation; Temp: temperature; FLDS: longwave radiation; VPD: vapor pressure deficit.

Secondly, we employed Long Short-Term Memory (LSTM) Networks [21], a member of the Recurrent Neural Networks (RNNs) family which specializes in incorporating temporal information in time-series data, to replace the FFNN counterpart in the same analyses. Instead of merely utilizing the climate drivers in the day before, we let the LSTM candidate take into consideration the data within the past thirty days. Compared to FFNN, our LSTM model achieved a consistently lower mean square error and higher model-data correlation (the middle panels in Figure 1).

The last neural network variant we analyzed was the transformers [22], which was recently introduced as a state-of-the-art natural language comprehension model whose utility covers the areas of machine translation [24], question-answering [25], and various other tasks. Similar to the LSTMs, thirty days' worth of data was provided as the input. In our task, there were several cases that the transformers (the bottom panels in Figure 1) outperformed FFNN, such as it maintained a higher Pearson correlation when shortwave radiation data was absent, but overall it did not demonstrate superior performance over FFNN, let alone the even better-performing LSTM.

Given that the LSTM was the best candidate among the three, we then inspected multiple variants of the LSTM model to explore the natural temporal memory of the photosynthetic process. The photosynthetic activities as measured by GPP depend on the past history of environmental drivers, but that temporal dependency is not likely to extend infinitely to the past. For example, whether it rained a million years ago would minimally affect the GPP tomorrow. An LSTM variant that utilizes information within a reasonable temporal span is likely to achieve optimal prediction accuracy. On the other hand, incorporating information over a period much shorter or longer than the natural memory duration is likely to yield sub-optimal results. The natural length of memory may be deduced by the performance of different LSTM variants, and it may shed light on the internal linkage of photosynthesis process across time.

Finally, we experimented on the forecasting ability of the LSTM models such that all variants were given the historical data on climate drivers over the experimentally determined two-week period, but were asked to predict the GPP on varying days ahead in the future. We would like to find out the turning point at which the model was no longer able to yield faithful predictions, which may again imply the temporal memory of photosynthetic dynamics. The results are displayed in Figure 2.

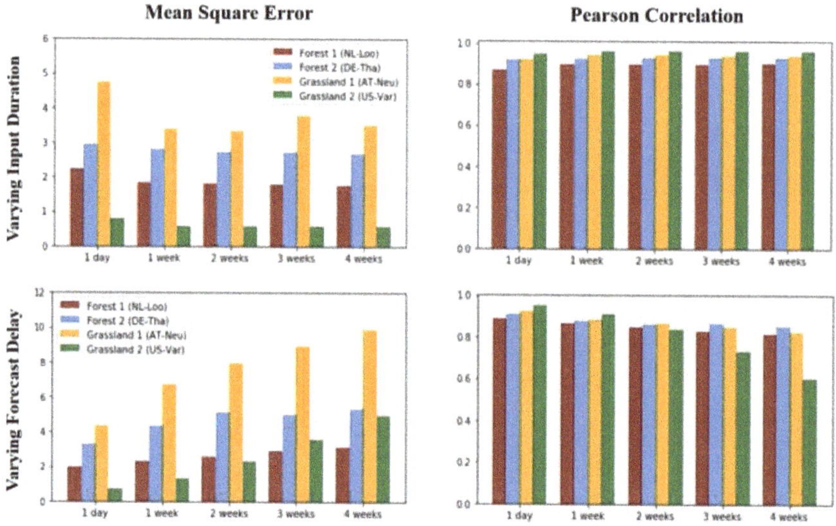

Figure 2. The past-history dependency and GPP forecasting ability of the LSTM model. Upper panel: predicting the current GPP using varying length of historic data on climate drivers. Lower panel: predicting the GPP on varying days ahead in the future using the same two-week length of historic data.

3. Discussion

In all three network candidates, the prediction accuracy were fairly consistent across multiple sites when all climate drivers were provided as the input (Figure 1). The second grassland site seemed to be the easiest to predict, judging from the fact that all three candidates achieved the best performance on that site, as indicated by low mean square error and high Pearson correlation. The reason behind this observation is yet to be discussed.

Furthermore, our leave-one-out experiment helped us identify the most influential climate drivers on GPP prediction. The accurate estimation of GPP relies heavily on knowing the shortwave radiation in the region. This observation is reasonable, as shortwave radiation is radiation energy with wavelengths around the range of visible light—the primary energy source for photosynthesis. Apart from shortwave, knowledge of any other climate driver does not significantly influence the GPP prediction accuracy consistently across all three neural network candidates, and interestingly enough this is also the case for GPP itself: presence or absence of past GPP data does not significantly affect the prediction of future GPP levels.

After identification of the most influential climate driver, the phenomenon that the second grassland is the "most predictable" site becomes more explainable. If we look ahead into the shortwave radiation distribution across the four sites shown in Figure 3 in the upcoming Materials and Methods section, we could draw a plausible conclusion that the abundance of shortwave radiation might be the reason behind the high predictability. The shortwave radiation distribution in the second grassland implies a lot of non-zero, information-carrying observations of this climate driver, which would reasonably improve the prediction accuracy of GPP.

In both the initial study and the leave-one-out experiments, the performance of the LSTM models was consistently better than the FFNN counterparts, which we would attribute to the inclusion of information over a longer duration of past history. The less appreciable performance of transformers compared to LSTMs was likely due to the fact that transformers were originally designed for natural language translation rather than for time-series interpretation. Besides being less task-specific, the transformer we implemented had more parameters than the other two candidates, meaning that it would require more data to optimize. That also might have yielded the less competitive performance given the same amount of training data.

In the study of past-history dependency and forecasting ability of the LSTM models (Figure 2), our experiment implied that photosynthesis has an internal memory length of two weeks. In the past-history dependency study, the LSTM performance peaked when the historical data fed into it covered the past two weeks; providing a more distant history of the climate drivers no longer improved the performance. In the forecasting study, our LSTM model was able to predict GPP two weeks into the future at most. Attempts to predict beyond two weeks demonstrated a significant decrease in prediction accuracy as indicated by higher mean squared error and lower correlation between the prediction and the respective ground truth. Such deterioration was especially obvious in grassland sites (orange and green bars). Our results on the optimal historical system memory were quite consistent with the longest temporal predictability, which were both about two weeks.

4. Materials and Methods

4.1. FLUXNET Dataset

We used Eddy covariance data collected from four different sites in the FLUXNET observational network [5]. Eddy covariance [26] is a measurement technique that quantifies ecosystem gas exchange and areal emission rates. By simultaneously measuring the velocity of the swirling wind with anemometers and the gas concentrations with infrared gas analyzers, the technique can eventually generate estimations of the flux of gases into or out of the ecosystem. By now, hundreds of sites are operating on a long-term and continuous basis. In this study, we included two evergreen needleleaf forest sites and two grassland sites (Table 1). Figure 3 shows the probability density distributions of the climate drivers at the four selected sites.

Table 1. Summary of the geographic location and basic information of the four FLUXNET sites included in this study. Time-varying data is represented in mean ± standard deviation. AP: annual precipitation. GPP: gross primary production, a measure that correlates to the rate of photosynthesis in a region.

Name	Latitude (N)	Longitude (E)	Elevation (m)	Land Cover Type	AP (mm yr^{-1})	Temp (°C)	GPP (gC m^{-2}day^{-1})
NL-Loo	52.16	5.74	25	Evergreen Needleleaf Forest	419 ± 829	10.1 ± 6.4	4.3 ± 3.1
DE-Tha	50.96	13.56	385	Evergreen Needleleaf Forest	420 ± 988	8.8 ± 7.9	5.1 ± 4.1
AT-Neu	47.11	11.31	970	Grassland	334 ± 814	6.8 ± 8.2	5.9 ± 5.9
US-Var	38.41	−120.95	129	Grassland	282 ± 980	15.8 ± 6.8	1.8 ± 2.8

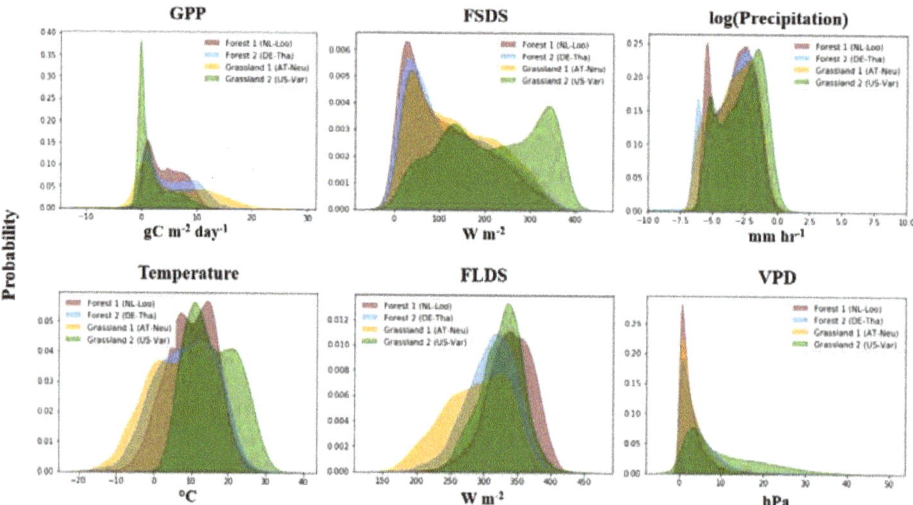

Figure 3. Probability densities of the observed climate drivers at the four FLUXNET sites included in this study. Red and orange are grassland sites while blue and green are evergreen needleleaf forest sites.

4.2. Neural Network Architectures and Formulas

4.2.1. FFNN

The first neural network candidate we explored, FFNN, composes of five layers of cells as shown in Figure 4b. Each cell is connected to all cells in the previous and the next layer but intact from other cells. Information from the input feature vectors, which in our case are the six climate drivers, can only pass from a shallower layer to a deeper layer but not the other way around, hence the "feedforward" part in the name of the model. Figure 4a indicates the composition of the input feature vectors and prediction ground truths for not only the FFNN but also the two other candidates.

An arbitrary subset of the FFNN, with n pre-synaptic cells attaching to the same post-synaptic cell, is depicted in Figure 5a for a detailed demonstration of the network. Conceptually, the signal passed to the post-synaptic cell is a weighted sum of the outputs from the pre-synaptic cells, and that signal goes through an activation function before it passes on to the cells in the next layer.

The mathematical description of this scenario is denoted in Equation (1).

$$y = f(\sum_{i=1}^{n} x_i \cdot w_i) \tag{1}$$

where x_i are the input features; w_i are neural network parameters, or in technical terms, weights, for each cell; f is the activation function, which is our implementation is a rectified linear unit (ReLU) [27]; and y_i is the output. For a cell in the first layer, the input features are the climate drivers, whereas for other cells deeper down the network, the input features are abstractions extracted from previous cells. For the single cell in the last layer, its output is the predicted GPP.

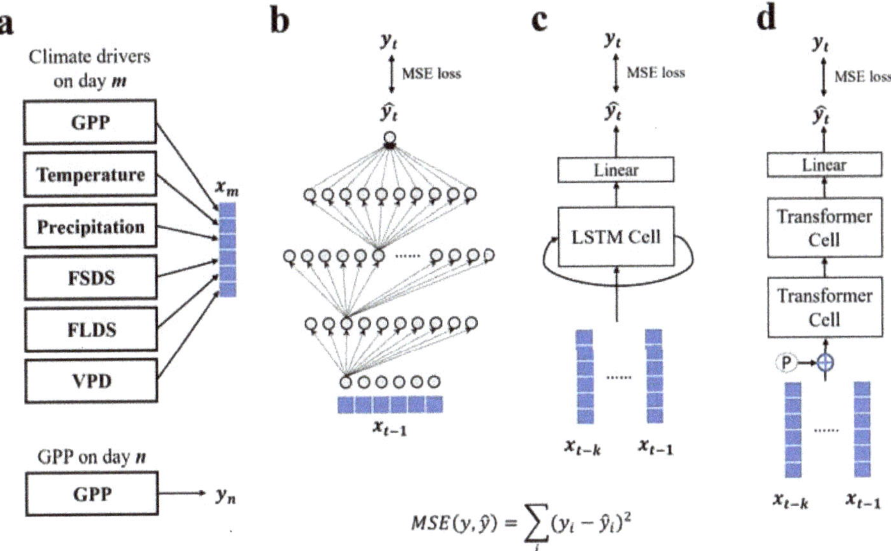

Figure 4. Complete depiction of the three neural network candidates used in our study. (**a**) The feature vectors and ground truths. The candidates take in the feature vectors as their inputs and generate predictions to approximate the ground truths. (**b**) Feedforward neural network (FFNN). Note: In each of the first three layers, only the connection from one chosen cell is drawn to avoid overcrowding the graph. In fact, every cell is connected to all cells in the adjacent layers. (**c**) Long short-term memory (LSTM). (**d**) Transformer. The symbols used are described in the next figure.

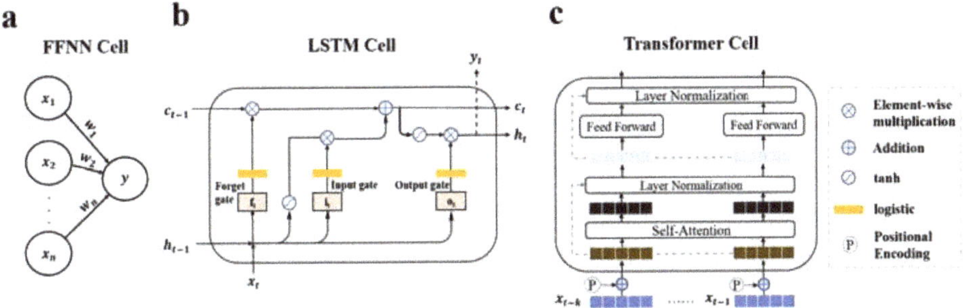

Figure 5. Illustration of the building blocks of the three neural network candidates used in our study. Symbols in the graphs are summarized in the panel on the right. (**a**) A representative subset of a FFNN consisting of multiple pre-synaptic cells and one post-synaptic cell. (**b**) A single cell of an LSTM. (**c**) A single block of a transformer.

4.2.2. LSTM

The second candidate, LSTM, models GPP based on not only climate drivers on the previous day but also climate drivers from the more distant past. As illustrated in Figure 4c, the six climate drivers over the past k days are sequentially fed into the LSTM in the correct chronological order. The order at which they enter the LSTM matters to the model, as the LSTM would adapt its states depending on the past and present information. On a conceptual level, LSTM shares some similar design philosophies with a hidden Markov model who computes the conditional probability $P(x_t | x_{t-k}, ..., x_{t-1}, state)$. The linear layer immediately after the LSTM cell in the figure stands for a feedforward layer that very much resembles the last layer in the FFNN counterpart: all the extracted features from the LSTM cell are condensed into a single FFNN post-synaptic cell whose output is the predicted GPP at day t.

The building block, or an LSTM cell, is demonstrated in Figure 5b. Conceptually, an LSTM cell keeps track of two internal states (i.e., **cell state** and **hidden state** which respectively corresponds to long-term and short-term memory) and consists of three gates (i.e., **forget gate** that determines what fraction of the previous cell state shall be maintained, **input gate** that decides how much new information from the current input shall be integrated to the cell state and hidden state, and **output gate** that updates the hidden state).

Mathematically speaking, given an input sequence $x = (x_1, ..., x_T)$, a standard LSTM computes the hidden vector sequence $h = (h_1, ..., h_T)$, which can be used as the output of the model, by iterating the following equations from $t = 1$ to T:

$$f_t = \sigma(W_f x_t + U_f h_{t-1} + b_f) \tag{2}$$

$$i_t = \sigma(W_i x_t + U_i h_{t-1} + b_i) \tag{3}$$

$$o_t = \sigma(W_o x_t + U_o h_{t-1} + b_o) \tag{4}$$

$$c_t = f_t \circ c_{t-1} + i_t \circ tanh(W_c x_t + U_c h_{t-1} + b_c) \tag{5}$$

$$h_t = o_t \circ tanh(c_t), \tag{6}$$

where σ is the sigmoid function, and f, i, o, c and h are respectively the input gate, forget gate, output gate, cell state and hidden state vectors. b_f, b_i, b_o and b_c are the biases (i.e., additive scalars) to their respective gates. Matrices W and U contain the model weights at the input and recurrent connections. x is the model input at each time step. The symbol "$\circ$" represents the point-wise multiplication operation.

4.2.3. Transformer

The third candidate, the transformer, also incorporates information from multiple time points in past history. Even though transformers and LSTMs may share similar high-level structures (Figure 4c,d), there exists a major difference between them. Transformers completely remove the concept of recurrence, or in other words, the way that LSTMs comprehend data as an ordered sequence that incrementally builds up information. They instead rely on two methods called positional encoding and self-attention to handle long series of data. The former means that although there is no longer a concept of "who comes before whom and after whom", the position of each data point is nevertheless encoded so that transformers do have spatial awareness. The latter method helps transformers to figure out "which data points" (or specifically, data points from "which positions") are more important and "pay more attention" to them, so that they can manage long sequence of data.

The fundamental building block of a transformer is shown in Figure 5c. All components should already be self-explanatory, except for the layer normalization blocks. The purpose of such blocks is to re-scale the output values from the previous layer in order to balance the data, prevent the model from being overwhelmed by extreme outliers, and prevent such outliers from propagating and deteriorating progressively.

4.3. Neural Network Implementation Details

Data was divided into train, validation and test sets at the ratio of 8:1:1, where the training set was used to train the model by updating model parameters according to the loss; the validation set was left for sanity checking each time the entire training set has been exhausted; and the test set was held out for the quantitative evaluation that yielded the results in Figures 1 and 2.

For all three candidates, model parameters were initialized as zeros and were iteratively optimized by back-propagation of loss (i.e., amount of difference between the model prediction and the GPP ground truth as determined by a loss function) via gradient descent. The loss function implemented for all candidates was mean square error loss (MSE loss). Adaptive learning rate (i.e., adjust model parameters at a larger magnitude at early stages of training, and gradually decrease that magnitude when the model reaches the fine-tuning stage) and early stopping (i.e., stop training when the model performance on the validation set no longer improves) were both implemented to prevent over-fitting (i.e., the phenomenon that the model performs great on the training set, most likely by "memorizing" the answers, but fails to generalize on the test set) and improve performance on the test set. Random seed was controlled to secure reproducibility: for example, the pseudo-randomized order at which each training sample was visited would be kept the same each time we ran the experiments.

Other more technical and less inspirational hyper-parameters, such as learning rate, batch size, etc., will not be mentioned in this paper, but can be found in the GitHub repository https://github. com/RosalieZhu/DL-CO2.

5. Conclusions

Greenhouse gas emissions such CO_2 could dramatically warm up climate system via positive radiative forcing effect. Fortunately, terrestrial ecosystems are able to mitigate the anthropogenic CO_2 emissions via photosynthesis. In this study, we aim to model the dynamics of plant photosynthesis activity using advanced machine learning frameworks with temporal awareness. Our results showed that incorporation of past information is important for improving model predictability. Our memory-based neural network model was able to successfully capture the temporal dynamics of plant photosynthesis at all four sites of interest, and it identified shortwave radiation as the most informative climate driver for faithful prediction of photosynthetic rates. Our modeling experiment also demonstrated a two-week internal system memory of the photosynthesis machinery.

Our study demonstrated that LSTM, with its capability of capturing temporal dependencies, is a technique worth investigating when researches study time series data regarding photosynthetic activities, or even environmental data in general. In the future, similar methods can be directly applied to evaluate other important climate drivers besides GPP, such as ecosystem respiration and net ecosystem exchange. Specifically, a more comprehensive study on mapping photosynthesis internal memory can be conducted if more data from additional FLUXNET sites that cover all land types are incorporated, and ultimately an internal memory map can be derived based on the prediction of the proposed LSTM network.

Author Contributions: Conceptualization, N.Z.; methodology, N.Z.; software, N.Z. and C.L.; validation, N.Z. and C.L.; formal analysis, N.Z. and C.L.; investigation, N.Z. and C.L.; resources, A.F.L. and J.G.; data curation, N.Z.; writing—original draft preparation, N.Z. and C.L.; writing—review and editing, N.Z., C.L., A.F.L. and J.G.; visualization, N.Z. and C.L. All authors have read and agreed to the published version of the manuscript.

Funding: This research received no external funding.

Acknowledgments: This study is conducted with the aid of the computational resources at the Mortimer B. Zuckerman Mind Brain Behavior Institute.

Conflicts of Interest: The authors declare no conflict of interest.

Abbreviations

The following abbreviations are used in this manuscript:

ANOVA	analysis of variance
ATP	adenosine triphosphate
CO_2	carbon dioxide
FFNN	feedforward neural network
FLDS	longwave radiation
FSDS	shortwave/solar radiation
GPP	gross primary product
HMM	hidden Markov model
HSD	honestly significant difference
LSTM	long short-term memory
MSE	mean square error
NADPH	nicotinamide adenine dinucleotide phosphate hydrogen
Pre	precipitation
RNN	recurrent neural network
RuBisCO enzyme	Ribulose-1,5-bisphosphate carboxylase oxygenase
VPD	vapor pressure defict

References

1. Stips, A.; Macias, D.; Coughlan, C.; Garcia-Gorriz, E.; San Liang, X. On the causal structure between CO_2 and global temperature. *Sci. Rep.* **2016**, *6*, 21691. [CrossRef] [PubMed]
2. Falkowski, P.; Scholes, R.; Boyle, E.; Canadell, J.; Canfield, D.; Elser, J.; Gruber, N.; Hibbard, K.; Högberg, P.; Linder, S.; et al. The global carbon cycle: A test of our knowledge of earth as a system. *Science* **2000**, *290*, 291–296. [CrossRef] [PubMed]
3. Beer, C.; Reichstein, M.; Tomelleri, E.; Ciais, P.; Jung, M.; Carvalhais, N.; Rödenbeck, C.; Arain, M.A.; Baldocchi, D.; Bonan, G.B.; et al. Terrestrial gross carbon dioxide uptake: Global distribution and covariation with climate. *Science* **2010**, *329*, 834–838. [CrossRef] [PubMed]
4. Friedlingstein, P.; Houghton, R.; Marland, G.; Hackler, J.; Boden, T.A.; Conway, T.; Canadell, J.; Raupach, M.; Ciais, P.; Le Quéré, C. Update on CO_2 emissions. *Nat. Geosci.* **2010**, *3*, 811–812. [CrossRef]
5. Baldocchi, D.; Falge, E.; Gu, L.; Olson, R.; Hollinger, D.; Running, S.; Anthoni, P.; Bernhofer, C.; Davis, K.; Evans, R.; et al. FLUXNET: A new tool to study the temporal and spatial variability of ecosystem-scale carbon dioxide, water vapor, and energy flux densities. *Bull. Am. Meteorol. Soc.* **2001**, *82*, 2415–2434. [CrossRef]
6. Medlyn, B.E. Physiological basis of the light use efficiency model. *Tree Physiol.* **1998**, *18*, 167–176. [CrossRef] [PubMed]
7. Farquhar, G.D.; von Caemmerer, S.V.; Berry, J.A. A biochemical model of photosynthetic CO_2 assimilation in leaves of C3 species. *Planta* **1980**, *149*, 78–90. [CrossRef]
8. Papale, D.; Valentini, R. A new assessment of European forests carbon exchanges by eddy fluxes and artificial neural network spatialization. *Glob. Chang. Biol.* **2003**, *9*, 525–535. [CrossRef]
9. Tramontana, G.; Ichii, K.; Camps-Valls, G.; Tomelleri, E.; Papale, D. Uncertainty analysis of gross primary production upscaling using Random Forests, remote sensing and eddy covariance data. *Remote Sens. Environ.* **2015**, *168*, 360–373. [CrossRef]
10. Jung, M.; Reichstein, M.; Bondeau, A. Towards global empirical upscaling of FLUXNET eddy covariance observations: Validation of a model tree ensemble approach using a biosphere model. *Biogeosciences* **2009**, *6*, 2001–2013. [CrossRef]
11. Xiao, J.; Zhuang, Q.; Baldocchi, D.D.; Law, B.E.; Richardson, A.D.; Chen, J.; Oren, R.; Starr, G.; Noormets, A.; Ma, S.; et al. Estimation of net ecosystem carbon exchange for the conterminous United States by combining MODIS and AmeriFlux data. *Agric. Forest Meteorol.* **2008**, *148*, 1827–1847. [CrossRef]

12. Xiao, J.; Zhuang, Q.; Law, B.E.; Chen, J.; Baldocchi, D.D.; Cook, D.R.; Oren, R.; Richardson, A.D.; Wharton, S.; Ma, S.; et al. A continuous measure of gross primary production for the conterminous United States derived from MODIS and AmeriFlux data. *Remote Sens. Environ.* **2010**, *114*, 576–591. [CrossRef]

13. Yang, F.; White, M.A.; Michaelis, A.R.; Ichii, K.; Hashimoto, H.; Votava, P.; Zhu, A.X.; Nemani, R.R. Prediction of continental-scale evapotranspiration by combining MODIS and AmeriFlux data through support vector machine. *IEEE Trans. Geosci. Remote Sens.* **2006**, *44*, 3452–3461. [CrossRef]

14. Yang, F.; Ichii, K.; White, M.A.; Hashimoto, H.; Michaelis, A.R.; Votava, P.; Zhu, A.X.; Huete, A.; Running, S.W.; Nemani, R.R. Developing a continental-scale measure of gross primary production by combining MODIS and AmeriFlux data through Support Vector Machine approach. *Remote Sens. Environ.* **2007**, *110*, 109–122. [CrossRef]

15. Jung, M.; Reichstein, M.; Ciais, P.; Seneviratne, S.I.; Sheffield, J.; Goulden, M.L.; Bonan, G.; Cescatti, A.; Chen, J.; De Jeu, R.; et al. Recent decline in the global land evapotranspiration trend due to limited moisture supply. *Nature* **2010**, *467*, 951–954. [CrossRef]

16. Jung, M.; Reichstein, M.; Margolis, H.A.; Cescatti, A.; Richardson, A.D.; Arain, M.A.; Arneth, A.; Bernhofer, C.; Bonal, D.; Chen, J.; et al. Global patterns of land-atmosphere fluxes of carbon dioxide, latent heat, and sensible heat derived from eddy covariance, satellite, and meteorological observations. *J. Geophys. Res. Biogeosci.* **2011**, *116*. [CrossRef]

17. Kondo, M.; Ichii, K.; Takagi, H.; Sasakawa, M. Comparison of the data-driven top-down and bottom-up global terrestrial CO_2 exchanges: GOSAT CO_2 inversion and empirical eddy flux upscaling. *J. Geophys. Res. Biogeosci.* **2015**, *120*, 1226–1245. [CrossRef]

18. Schwalm, C.R.; Williams, C.A.; Schaefer, K.; Arneth, A.; Bonal, D.; Buchmann, N.; Chen, J.; Law, B.E.; Lindroth, A.; Luyssaert, S.; et al. Assimilation exceeds respiration sensitivity to drought: A FLUXNET synthesis. *Glob. Chang. Biol.* **2010**, *16*, 657–670. [CrossRef]

19. Schwalm, C.R.; Williams, C.A.; Schaefer, K.; Baldocchi, D.; Black, T.A.; Goldstein, A.H.; Law, B.E.; Oechel, W.C.; Scott, R.L. Reduction in carbon uptake during turn of the century drought in western North America. *Nat. Geosci.* **2012**, *5*, 551–556. [CrossRef]

20. Dou, X.; Yang, Y.; Luo, J. Estimating Forest Carbon Fluxes Using Machine Learning Techniques Based on Eddy Covariance Measurements. *Sustainability* **2018**, *10*, 203. [CrossRef]

21. Hochreiter, S.; Schmidhuber, J. Long short-term memory. *Neural Comput.* **1997**, *9*, 1735–1780. [CrossRef] [PubMed]

22. Vaswani, A.; Shazeer, N.; Parmar, N.; Uszkoreit, J.; Jones, L.; Gomez, A.N.; Kaiser, Ł.; Polosukhin, I. Attention is all you need. In *Advances in Neural Information Processing Systems*; The MIT Press: Cambridge, MA, USA, 2017; pp. 5998–6008.

23. Tukey, J.W. Comparing individual means in the analysis of variance. *Biometrics* **1949**, *5*, 99–114. [CrossRef] [PubMed]

24. Wang, Q.; Li, B.; Xiao, T.; Zhu, J.; Li, C.; Wong, D.F.; Chao, L.S. Learning deep transformer models for machine translation. *arXiv* **2019**, arXiv:1906.01787.

25. Devlin, J.; Chang, M.W.; Lee, K.; Toutanova, K. Bert: Pre-training of deep bidirectional transformers for language understanding. *arXiv* **2018**, arXiv:1810.04805.

26. Burba, G. *Eddy Covariance Method for Scientific, Industrial, Agricultural and Regulatory Applications: A Field Book on Measuring Ecosystem Gas Exchange and Areal Emission Rates*; LI-Cor Biosciences: Lincoln, NE, USA, 2013.

27. Nair, V.; Hinton, G.E. Rectified linear units improve restricted boltzmann machines. In Proceedings of the 27th International Conference on Machine Learning (ICML-10), Haifa, Israel, 21–24 June 2010; pp. 807–814.

Article

Crude Oil Prices Forecasting: An Approach of Using CEEMDAN-Based Multi-Layer Gated Recurrent Unit Networks

Hualing Lin * and Qiubi Sun

The Department of Statistics, School of Economics and Management, Fuzhou University, Fuzhou 350018, China; sqb@fzu.edu.cn
* Correspondence: m160710006@fzu.edu.cn

Received: 26 February 2020; Accepted: 23 March 2020; Published: 25 March 2020

Abstract: Accurate prediction of crude oil prices is meaningful for reducing firm risks, stabilizing commodity prices and maintaining national financial security. Wrong crude oil price forecasts can bring huge losses to governments, enterprises, investors and even cause economic and social instability. Many classic econometrics and computational approaches show good performance for the ordinary time series prediction tasks, but not satisfactory in crude oil price predictions. They ignore the characteristics of non-linearity and non-stationarity of crude oil prices data, which hinder an accurate prediction and eventually lead to poor accuracy or the wrong result. Empirical mode decomposition (EMD) and ensemble EMD (EEMD) solve the problems of non-stationary time series forecasting, but they also generate new problems of mode mixing and reconstruction errors. We propose a hybrid method that is combination of the complete ensemble empirical mode decomposition with adaptive noise (CEEMDAN) and multi-layer gated recurrent unit (ML-GRU) neural network to solve the abovementioned issues. This not only deals with the issue of mode mixing effectively, but also makes the reconstruction error of data close to zero. Multi-layer GRU has an excellent ability of nonlinear data-fitting. The experimental results of real WTI crude oil dataset show that the proposed approach perform better in crude oil prices forecasts than some state-of-the-art models.

Keywords: crude oil prices; forecasting; complete ensemble empirical mode decomposition with adaptive noise (CEEMDAN); multi-layer gated recurrent unit (ML-GRU)

---

1. Introduction

Crude oil was once considered to be the blood flowing through the veins of the world economy and played an extremely critical role in the development of the world economy. According to a report by the U.S. Energy Information Administration (EIA), although renewable energy production and consumption both reached their highest share in 2018, fossil fuel still accounted for 80% of the United States' energy consumption. In light of the International Energy Agency (IEA) data, the global consumption of oil reached 1.0075 million barrels per day in 2019. In meeting world energy needs, oil still plays the most important role. Asian emerging market countries have become the main contributors to the growth in crude oil demand. The rapid economic growth has prompted them to significantly increase demand for crude oil. For example, China's oil consumption has soared from an average of 69,700 barrels per day in 2005 to 145,100 barrels per day in 2019. As a production factor, the increase in crude oil prices will lead to an increase in the non-oil companies' production costs and a decrease in profits. Sadorsky [1] verified the impact of fluctuations in crude oil prices on companies of different firm size through evidence from the stock market. Rising oil prices may lead to inflation and hinder economic growth. Volatility in oil prices increases risk and uncertainty to financial markets [2]. Not only that, but the continued collapse or the sudden plunge in oil prices can also have a huge impact on the economic development and financial markets of oil-producing countries [3]. An obvious piece of evidence is the crisis signal coming from the credit default swap (CDS) market. The widening of the CDS spread means that weak crude oil prices have caused investors to worry about the fiscal sustainability of some oil producing countries [4]. The study of Haushalter et al. [5] concluded that there is a negative correlation between the price of crude oil and the debt ratio of oil producers.

Based on the above discussion, it is obviously of great significance to find a method that can accurately predict the price of crude oil. This means that policy makers will have more time to introduce countermeasures to achieve the goal of avoiding or reducing risks. However, the trend of oil prices is affected by not only the factors of market supply and demand, but also the other non-market factors, such as geopolitical, alternative forms of energy, recession, war, natural disasters, and technological development. Various uncertainties affect the crude oil market. The fluctuations in these factors cause the nonlinear, volatile, and chaotic tendency of crude oil prices. Therefore, achieving sufficiently reliable and accurate forecasting of crude oil prices has become one of the most challenging issues.

In the past decades, various techniques have been tried to forecast the trend of crude oil prices. The classic statistical or econometric models such as autoregressive integrated moving average (ARIMA) model and the autoregressive conditional heteroskedasticity (ARCH)/generalized ARCH (GARCH) family model are widely used in the time series prediction tasks [6–12]. Zhao and Wang [9] used the ARIMA model to model and predict crude oil prices based on the international crude oil prices data from the 1970s to 2006. Mohammadi and Lixian [13] applied the ARIMA-GARCH model to forecast the conditional mean and volatility of weekly crude oil spot prices in eleven international markets. Aamir and Shabri [14] used the Box-Jenkins ARIMA, GARCH and ARIMA-Kalman models to model and forecast the monthly crude oil prices in Pakistan. The premise of applying these classic models represented by ARMA /ARIMA is that there is an autocorrelation in the crude oil prices series. So, the historical data is used to infer the future prices of crude oil. These methods are suitable for capturing the linear relationships in time series analysis but not for the nonlinear time series [15]. In addition to the classical methods mentioned above, the survey forecasting method is another alternative method for crude oil price forecasting [16]. Kunze et al. [17] empirically studied the performance of survey-based predictions of crude oil prices. The evaluation shows that the prediction accuracy of the survey-based forecasts is not as good as that of the naive method in the short-term crude oil prediction, but with the rise of the prediction horizon, the accuracy of the former will exceed that of the naive method. This study demonstrates that survey predictions are not suitable for the short-term prediction of crude oil prices.

Owning to the drawbacks of the classic approaches and the special features of crude oil price data, artificial intelligence (AI), such as machine learning, deep learning or hybrid methods provide more excellent nonlinear data predictive performance. In term of forecasting crude oil prices, the approaches of AI are being widely used as alternative to the classic technologies. Li et al. [18] proposed an approach that incorporates ensemble empirical mode decomposition (EEMD), sparse Bayesian learning (SBL), for forecasting crude oil prices. Xie et al. [19] used support vector machines to forecast crude oil prices and compared the performance with ARIMA and BPNN's. Experiments show that SVM has better performance than the other two methods. Fan et al. [20] propose an independent component analysis based SVR scheme, for crude oil prices predictions. This approach starts from an independent component analysis to decompose crude oil prices series into independent components, which are respectively forecasted by support vector regression (SVR).

Crude oil prices trends have significant nonlinear and non-stationary characteristics. Regardless of the traditional statistics or machine learning methods, it is difficult to obtain satisfactory results by directly predicting the original crude oil prices series. Therefore, scholars usually adopt the method that, first, the original crude oil prices series is decomposed into multiple same time scales or relatively simple sub-sequences by a signal decomposition algorithm. Next, create multiple subtasks. Each sub-task completes the prediction of a sub-sequence individually. Finally, the results of all subtasks are linearly calculated and the forecasting is obtained. Common signal decomposition methods, such as the wavelet transform, EMD, and EEMD, have been widely used to process the non-stationary time series. Yu et al. [21] proposed an EMD-based neural network crude oil price forecasting. He et al. [22] come up with a wavelet-based ensemble model to improve the predictive accuracy of oil prices. This approach introduced the wavelet to generate dynamic basic data within a finer time-scale domain. Hamid and Shabri [23] proposed a wavelet multiple linear regression method in daily forecasts of crude oil price. Wu et al. [24] propose a model based on EEMD and long short-term memory (LSTM) for crude oil price forecasting. Zhou et al. [25] introduced a hybrid approach of complete ensemble empirical mode decomposition with adaptive noise (CEEMDAN) and XGBOOST-based approach to forecast crude oil prices.

Both wavelet analysis and EMD are becoming the common tools for analyzing non-stationary time series but have their own limitations. The wavelet method can't achieve an adaptive decomposition in the light of time scales. It has been proven above that EMD perform well in extracting signals from the non-stationary data. Mode mixing is the main limitations of EMD. It is a consequence of signal intermittency which could result in the physical meaning of individual intrinsic mode function (IMF) unclear [26]. To overcome this limitation of EMD, an ensemble EMD named EEMD, was subsequently introduced by Wu and Huang [27]. EEMD adds white noise to the original signal. After a sufficient number of EMD tests, the only lasting component of signal is then identified as the substantial answer. EEMD eliminates the effects of mode mixing of EMD, but still retains some noise in the IMFs, which affect the accuracy of signal reconstruction [28].

In summary, much effort has been made to improve the accuracy of forecasting crude oil prices, but a more effective approach should be developed. The goal of this study is to propose a new novel approach of CEEMDAN-based multi-layer gated recurrent unit networks (CEEMDAN-ML-GRU). CEEMDAN is a variant of EEMD. In the applying of CEEMDAN, the multiple groups of adaptive white noise is added to the original data at each stage of the decomposition and a unique residue is computed to obtain each mode. CEEMDAN is complete, with a numerically negligible error of signal reconstruction. Due to the excellent characteristics of CEEMDAN, in recent years, some researchers have tried to apply it to no-stationary time series analysis [29,30]. The gated recurrent unit (GRU), like LSTM, is a recurrent neural network with a gating mechanism, but it has fewer parameters than LSTM, as it lacks an output gate. GRU's performance on certain tasks was found to be similar or even better to that of LSTM. A GRU network with a multi-layer stack structure has more powerful performance than a single-layer structure. The following experiments show that our proposed hybrid model goes on better than some other state-of-the-art's in oil price forecast.

The content of this paper is organized as follows: Section 2 review the background works related to our method. Section 3 introduce the proposed method in detail, Section 4 applies the proposed approach to forecast crude oil prices of West Texas Intermediate (WTI), then compares it with other standard models and some state-of-the-art hybrid models. Finally, Section 5 concludes the study and summarizes several main interesting issues for future research.

2. Related Work

This section will briefly review the existing works that closely relate to our proposed approach.

2.1. Artificial Neural Networks (ANN)

ANN is made up of many "neurons", whose output can be the input of another neuron. One kind of classic ANN, the multilayer perceptron (MLP) is illustrated in Figure 1. In the graph illustrated in this figure, the variable X represents the input, and the circles represent the "neurons" of the network. This ANN has three layers which relative positions, arranged from left to right, are in sequence: input layer, output layer, and hidden layer. In Figure 1, the circles represent the nodes of the network.

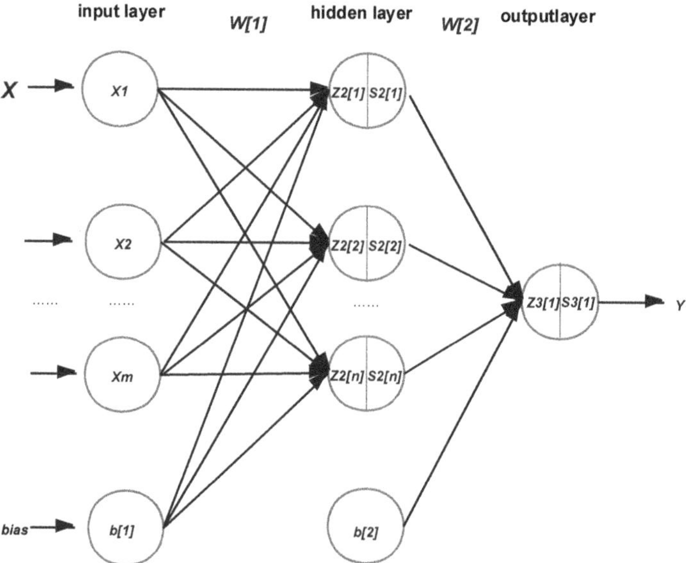

Figure 1. A multilayer perceptron (MLP) artificial neural network, which is a feedforward artificial neural network.

There are m input neurons (nodes), n hidden neurons (nodes) and one output neuron (node) in this network that each layer of neurons receives inputs from the previous layers. This type of network is also called a multi-layer feedforward network, where the output of a node in one layer is the input of the next layer. The nodes in the hidden layer receive the output of the previous layer and perform weighted linear combination of the inputs. The result is then modified by an activation function before being output. We introduce several concepts that are closely related to neural networks:

(1)   Activate Function, it can normalize the output to a given range to ensure that the model is convergent.   Each neuron accepts input and passes it through an activation function. The commonly used activation functions include Sigmoid, Tanh and ReLu.

(2)   Backpropagation, it is an algorithm commonly used to train the neural networks. After the inputs are loaded into the network, they pass forward through the neural network. Given an initial weight, the network provides an output for each neuron. When there is an error between the classification or regression results and the observations, the back-propagation mechanism comes into play. It helps to adjust the weights of the neurons, bringing the results closer and closer to the known true results.

(3)   Optimization algorithm, they are commonly used mathematical techniques in neural network optimization, and use the backpropagation to calculate the gradients. An example is gradient descent, the most common optimization method. In each training cycle, the best strategy for parameter (weight) adjustment is determined by observing the derivative of the error function with respect to each parameter. It enables the parameters to be updated in the negative gradient direction of the error function during each training to achieve the purpose of minimizing the error. In deep learning, the commonly used optimizers include stochastic gradient descent (SGD), adaptive gradient algorithm (Adagrad), adaptive moment estimation (Adam), etc.

2.2. Backpropagation

In ANN, especially in deep learning, backpropagation is an algorithm widely used to train feed-forward neural networks for supervised learning. The backpropagation algorithm is to compute the gradient of the loss function with respect to each parameter by the chain rules, and iterate backward from the last layer to avoid repeating calculation of the intermediate term of the chain. All of deep learning models mentioned later, such as recurrent neural network (RNN), LSTM and GRU use the backpropagation for training the network. The MLP is also shown in Figure 1.

Given a training set of m examples. The backpropagation algorithm is as follows:

(1) Perform a forward propagation, and so on up to the output layer.

$$Z_i^{l+1} = \sum_{j=1}^{m} W_{ij}^{(l)} x_j + b_i^{(l)} \tag{1}$$

$$S_i^{(l+1)} = \sigma\left(Z_i^{l+1}\right) = \sigma\left(\sum_{j=1}^{m} W_{ij}^{(l)} x_j + b_i^{(l)}\right) \tag{2}$$

$$F_{W,b}^{(l+1)}(X) = \sigma\left(\sum_{i=1}^{n} S_i^{(l+1)} + b_i\right). \tag{3}$$

The (W,b) is the parameter where $W_{ij}^{(l)}$ denote the weight associated with the connection between unit i in layer l, and unit j in layer $l+1$. $b_i^{(l)}$ is the bias associated with unit i in layer l. σ denotes the sigmoid function, which can transform the data into a value in the range of 0–1, thereby serving as a gate signal. $S_i^{(l)}$ denote the activation of unit i in layer l. The computation steps of the neural network Figure 1 represent is given by:

$$S_1^{(2)} = \sigma\left(W_{11}^{(1)} x_1 + W_{12}^{(1)} x_2 + W_{13}^{(1)} x_3 + b_1^{(1)}\right)$$
$$S_2^{(2)} = \sigma\left(W_{21}^{(1)} x_1 + W_{22}^{(1)} x_2 + W_{23}^{(1)} x_3 + b_2^{(1)}\right)$$
$$S_3^{(2)} = \sigma\left(W_{31}^{(1)} x_1 + W_{32}^{(1)} x_2 + W_{33}^{(1)} x_3 + b_3^{(1)}\right)$$
$$F_{W,b}(X) = S_1^{(3)} = \sigma\left(W_{11}^{(2)} S_1^{(2)} + W_{12}^{(2)} S_2^{(2)} + W_{13}^{(2)} S_3^{(2)} + b_1^{(2)}\right)$$

(2) Define the overall cost function to be:

$$C_{W,b} = \frac{1}{m} \sum_{i=1}^{m} \left(\frac{1}{2}\left\|F_{W,b}(X) - Y\right\|^2\right) \tag{4}$$

This cost function is used to compute error between the actual output and the expected output.

(3) Compute the partial derivatives with respect to (W,b):

$$\delta_j^L = \frac{\partial C_{W,b}}{\partial S_i^{(l)}} \sigma'\left(\sum_{i=1}^{n} S_i^{(l+1)} + b_i\right) \tag{5}$$

The derivative tells us the direction of movement of the weight values and how to get a lower cost in the next iteration.

2.3. Gradient Descent (GD)

In neural networks or deep learning, GD is one of the most common optimization algorithms used to minimize the loss function by iteratively moving in the steepest direction. The process of updating the parameters (W,b) in one iteration of GD is as shown below:

(1) Initialize the $\Delta W^{(l)}$ and $\Delta b^{(l)}$.
(2) Use backpropagation to compute $\nabla W^{(l)} C_{W,b}$ and $\nabla b^{(l)} C_{W,b}$.

(3)   Set: $\Delta W^{(l)} := \Delta W^{(l)} + \nabla_{W^{(l)}} C_{(W,b;x,y)}$, $\Delta b^{(l)} := \Delta b^{(l)} + \nabla_{b^{(l)}} C_{(W,b;x,y)}$

(4)   Update the parameters:

$$W_{ij}^{(l)} = W_{ij}^{(l)} - \alpha \frac{\partial}{\partial W_{ij}^{(l)}} C_{(W,b)} \tag{6}$$

$$b_i^{(l)} = b_i^{(l)} - \alpha \frac{\partial}{\partial b_i^{(l)}} C_{(W,b)} \tag{7}$$

where α is the learning rate, which specifies how aggressively the gradient descent should jump between successive iteration.

3. Methodology

This section will not only detail the method proposed in this paper, but also introduce some models that are closely related to the proposed method. They include signal processing algorithm such as EMD, EEMD and CEEMD, recurrent neural networks such as RNN, GRU and ML-GRU.

3.1. EMD, EEMD and CEEMDAN

Through EMD, the nonlinear and non-stationary signal can be adaptively decomposed into the limited IMFs based on local characteristic of time scales. The essence of this method is to eliminates the interference of noise and identify the intrinsic oscillatory modes in the data empirically. To obtain the valuable instantaneous frequencies, the IMFs must satisfy two conditions:

(1)   The numbers of extremes and zero crossings of the sequence must be equal or differ by no more than one.
(2)   At any location, the mean of the envelope determined by the local extrema is zero [26].

The EMD is developed as follows:
Connect all the local maxima (minima) with a cubic spline as the upper (lower) envelope.
Get the first IMF by calculating the difference between the original data and local mean envelope:

$$M(t) = \frac{1}{2}[U(t) - L(t)] \tag{8}$$

$$\text{IMF}(t) = X(t) - M(t) \tag{9}$$

where $U(t)$ and $L(t)$ are the upper envelope and the lower envelope respectively. If the difference original data $X(t)$ to the mean envelope $(M(t)$ meets the IMF constraints, this difference is the new IMF.

The result of subtracting all the previous IMFs from original data is the current residue. Using the residue as the new input and repeating the above procedure, the next IMF can be obtained:

$$R_K(t) = X(t) - \sum_{i=1}^{K} \text{IMF}_i \tag{10}$$

$$\text{IMF}_{K+1}(t) = R_K(t) - \text{IMF}_{K+1} \tag{11}$$

where $R_K(t)$ is the residue series after K-th decomposition. A complete decomposition process stops when the residue, $R_K(t)$, has been a monotonic function.

There are some drawbacks in EMD, mainly as follows: (a) In IMFs mode mixing exists. It means that the IMFs composed of oscillations of different time-scales and no longer have physical meaning. (b) The effects of end affect the results of decomposition. To overcome the drawbacks, Huang et al. [27] introduced a novel EEMD approach that utilizes the advantages of white noise to eliminate the effects of mode mixing. After EEMD, the reconstructed data includes residual noise and disparate parameters of noise can produce disparate number of modes [28].

CEEMD, proposed by Torres et al. [28]. The final decomposition result is obtained by taking the average by adaptively adding white noise to the time series with the same magnitude and opposite direction. After engaging in EMD processing, the output of CEEMDAN obeys a Gaussian distribution. Compared with EEMD and CEEMD, this method effectively eliminates the problem of mode mixing, and no matter how many times the decomposition, the reconstruction error of the signal is almost zero, the completeness is better, and at the same time solves the problem of low decomposition efficiency, greatly reducing the calculation cost. CEEMDAN's spectra shows a more accurate decomposition of the frequency than the EEMD's. The same time series, the number of iterations of CEEMDAN shifting is usually equivalent to half of the EEMD's.

The implementation of the algorithm is summarized as follows:

Generate a white noise plus series $(x_i(t) = x(t) + w_i(t))$. Here, $w_i(t)$ denote the white noise of finite variance, while $x(t)$ represent the original data. Then decompose $x_i(t)$ to obtain the IMFs:

$$\overline{imf_i} = \frac{1}{n}\sum_{j=1}^{n} imf_i^j(t) \tag{12}$$

where $w_i^j(t)$ is the white noise added at the j-th time with the mean equal to zero and variance equal to one. The $\overline{imf_i}$ is the i-th mode component obtained after the signal is decomposed by CEEMDAN.

Calculate the first residue:

$$r_1(t) = x(t) - \overline{imf_1} \tag{13}$$

Decompose the residue to get the second IMF and calculate the second residue:

$$\overline{imf_2} = \frac{1}{n}\sum_{j=1}^{n} E_1\{r_1(t) + \varepsilon_1 E_1[w_2^j(t)]\} \tag{14}$$

$$r_2(t) = r_1(t) - \overline{imf_2} \tag{15}$$

Decompose the $(k-1)$th residue of $r_{k-1}(t)$ and extract the $\overline{imf_k}(t)$. The process can be demonstrated in the following equation ($k = 2, 3, \dots K$):

$$\overline{imf_k(t)} = \frac{1}{n}\sum_{j=1}^{n} E_1\left(r_{k-1}(t) + w_{k-1}^j E_{k-1}\left(\varepsilon^j(t)\right)\right) \tag{16}$$

Calculate the k-th residue:

$$r_k(t) = x(t) - \sum_{i=1}^{k} imf_i \tag{17}$$

where k indicates the number of IMFs. The attributes of the original time series are denoted by all the IMFs exacted from different time-scales. The only residue demonstrates the trend, which is smoother than that of the original time series.

3.2. GRU and Multi-LayerML-GRU

3.2.1. RNN and GRU

Among deep learning methods, RNN is a powerful method for processing time series data. It is widely used in many fields such as finance [31,32], industry and engineering [33], machine translation [34], speech recognition [35], economic prediction [36], and so on. As shown in Figure 2, RNN has certain information persistence capabilities, which enable information to be passed from one time-step to the next. However, the classic RNN does not have the ability to store and memorize data for a long time, so that it cannot capture long-term historical information. In addition, in the reverse process of model training, once the sequence is too long, the RNN will cause the problem of gradient explosion or gradient disappearance. To overcome these drawbacks, Hochreiter and Schmidhuber [37] proposed the LSTM neural network which is capable of learning long-term dependencies of time series, as well as forgetting the worthless information based on the current input. LSTM has since replaced RNN as the most widely used recurrent neural network. LSTM has four gated unit that capable adaptively regulate the information flow inside the unit. An LSTM neural network usually requires many gated units, which need train a large number of parameters and occupy more computing resources. In order to reduce the training parameters and simply the neural network, Cho et al. [38] proposed a neural network named GRU which only have two gated units in the hidden unit.

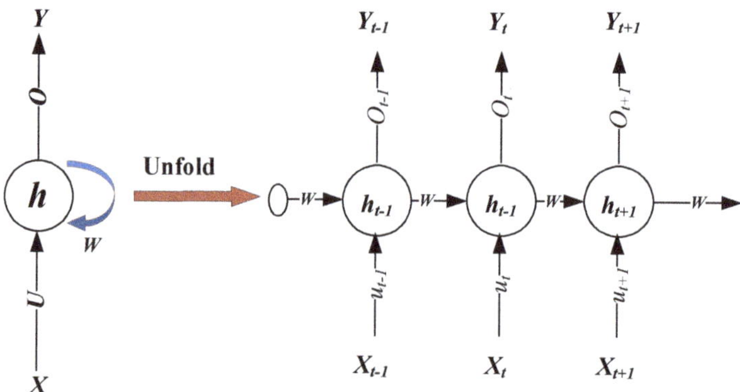

Figure 2. Workflow graph of RNN. X denotes current input of neural network; Y denotes the output of neural network; W, U and O are the parameters of recurrent neural network; h is the neuron state of the hidden layer; The state at time t is related to the current input X and the hidden at time $t-1$.

As one of the variants of RNN, the input and output structure of GRU neural network is the same as that of RNN and LSTM, as shown in the Figure 3 The neurons of GRU receive the hidden state (h_{t-1}) of the neuron of the previous neuron, and the current input x_t. After passing through the gating unit, the neural network gets the output y_t and passes the hidden state h_t to the next neuron.

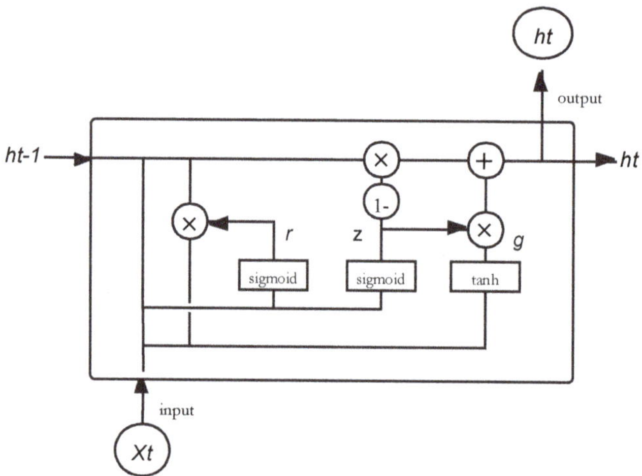

Figure 3. The internal structure of the GRU cell. r_t —Reset gate. It allows the cell to forget certain parts of the state; z_t—Update gate.

The GRU architecture, like LSTM, learns the long-term dependence of time series based on a gate mechanism that includes reset gate and update gate. The former is used to control how much information in the previous state will be ignored. The latter is used to control how much information from the historical state is brought into the current state. GRU neural networks, like other neural networks, consist of a large number of basic neurons. They are interconnected in a complex network. A single neuron cell is shown in Figure 2. The specific process is designed as follows:

$$z_t = \sigma(w_{xz}x_t + w_{hz}h_{t-1}) \tag{18}$$

$$r_t = \sigma(w_{xr}x_t + w_{hr}h_{t-1}) \tag{19}$$

The internal structure of the GRU cell is shown in Figure 3.

The update gate is used to control the extent to which the state information from the previous moment is retained to the current state. The more the value of the update gate approaches 0, the more the state information from the previous moment is brought into the current state. The update gate signal is the closer to 0, the more data it remembers. The closer to 1, the less it is forgotten.

Begin with getting the gating signal, GRU gets the reset data through the gate; then combine it with the input x_t, and use a tanh activation function to shrink the data to the range from -1 to 1. The formula is shown in the following:

$$g_t = \tan h\left(w_{xg}x_t + w_{hg}(r_t \times h_{t-1})\right) \tag{20}$$

The last step of GRU, we can call it "update memory" phase. Combined with the previous discussion, this step forgets some of the dimensional information passed in and add some new information inputted by the current neuron to the state variable h_t:

$$h_t = (1 - z_t) \times h_{t-1} + z_t \times g_t \tag{21}$$

3.2.2. Multi-Layer GRU Architecture

If the problem is too complicated, a recurrent neural network with a single layer structure is not enough to abstract the problem, and a multi-layer neural network is a better alternative. The multi-layer neural network has more hidden layers and more powerful computing capabilities which is the key to solving complex problems. The proposed model is one kind of forward multi-layer neural networks, which demand the input layer are required to have the same number of input dimensions as the input vector. The architecture and workflow of the ML-GRU are shown in Figure 4. The output dimensions of the last layer demand only be equal to the number of labels for classification or a single value of prediction for regression. With each training, the parameters are continuously updated. This process continues until the output of the network is closer to the desired output.

The multi-layer architecture determines that data flows through more neurons and more parameters need to be trained, as well as more powerful than the single layer network. If the recurrent neural network is too deep beyond what is necessary, the computational cost will be expensive. Also, the phenomenon of overfitting may occur.

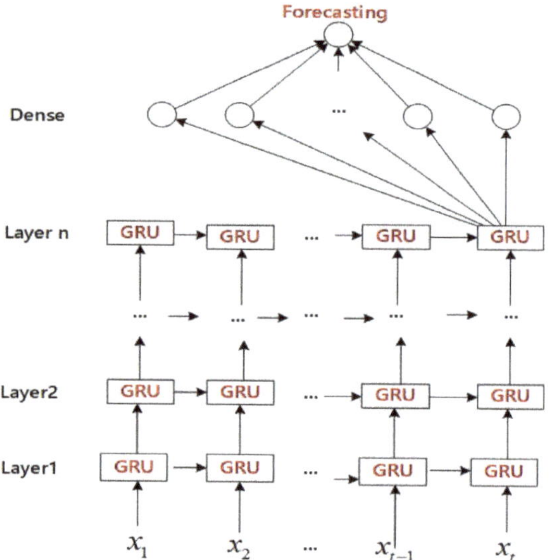

Figure 4. Architecture and workflow of ML-GRU.

3.3. CEEMDAN-Based Multi-Layer Gated Recurrent Unit Networks (CEEMDAN-ML-GRU)

In this study, we introduce a novel approach combining CEEMDAN and multi-layer GRU neural networks. We call this model CEEMDAN-ML-GRU for crude oil price forecasts. Figure 5 shows the architecture and workflow of the hybrid model. It aimed at improving the exiting crude oil price forecast techniques, which are less efficient or have poor accuracy in dealing with nonlinear and nonstationary regression tasks.

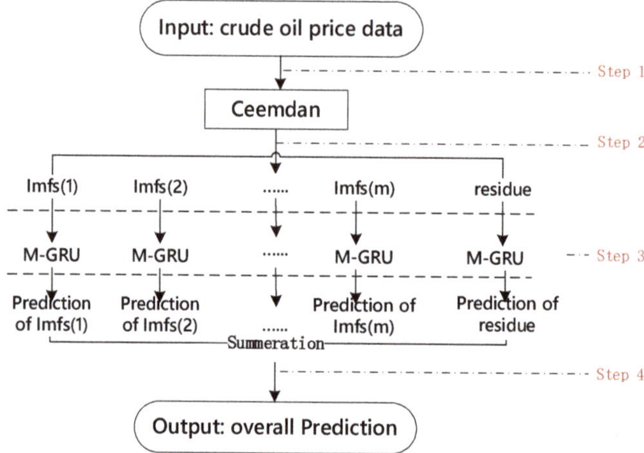

Figure 5. CEEMDAN-based Multi-layer Gated Recurrent Unit Networks (CEEMDAN-ML-GRU). This approach adopts a strategy of "divide and process", which transforms a complicated issue into Several simple issues, and processes them independently.

Firstly, CEEMD technology is used to add positive and negative paired white noise to the original exchange rate sequence, which overcomes the problem of large EEMD reconstruction errors and poor completeness of decomposition, and effectively improves the decomposition efficiency. Then the original exchange rate sequence is decomposed into IMFs based on different characteristics. Finally, the multi-layer LSTM based forecasting model is input for each component, and each of the IMF prediction results is superimposed to obtain the desired overall prediction.

As mentioned in the introduction, signal decomposition techniques such as wavelet transform, EMD and EEMD have been used for the analysis of time series of predicting energy prices. Machine learning, especially neural networks and deep learning methods, have also been applied to crude oil price forecasting to improve the learning process and prediction accuracy of crude oil price data. Signal decomposition technology is good at processing non-stationary data, and deep learning shows excellent performance when analyzing time series with nonlinear and long-term dependency characteristics. Some scholars have proved that integrating signal decomposition and deep learning methods for crude oil prices forecasting gained better results than only using a single method. In Section 1, we have mentioned that Wu et al. [24] proposed a novel model based on EEMD and long LSTM for crude oil price forecast. We will compare them experimentally with the proposed method in the next section. This kind of hybrid model is able to synthesize the strengths of each hybrid method, and significantly avoids the negative impact of the single method's inherent disadvantages on prediction performance.

The model first performs CEEMDAN on the original crude oil prices data. Then we input the previously decomposed IMF into the ML-GRU neural network. As for the output, it is meaningless to predict IMF alone, so we must use all the IMFs obtained by decomposing one sample at a time as the input of a single training or test to directly predict the price trend of crude oil. We use a one-step prediction strategy. By extracting the characteristics of IMFs, ML-GRU can use the crude oil prices trend data of the last p days to predict the price trend of the next day. CEEMDAN-ML-GRU prediction usually includes the following four main steps:

Step 1: Data preparing. In order to make the data meet the requirements of the model input, we first preprocess the original data, where involved data cleaning, data reduction and data transformation.

Step 2: Data decomposition. The training and test sets are decomposed into sets of IMFs and residue using the CEEMDAN method. The original complex time series $x(t)$, $t = 1, 2, \ldots, n$ is split into a training set and a test set in a supervised form.

Step 3: Model training. Input the IMFs of training set into Multi-layer GRU neural network for training.

Step 4: Price forecasting. Input the IMFs of test set into the trained multi-layer neural network to make one-step ahead forecasting for verification.

4. Experiments

4.1. Datasets

In this section, through a series of experiments, we verified the proposed CEEMDAN-ML-GRU model is more advanced than the state-of-the-art methods, which include the EEMD-LSTM. In order to measure and compare the forecasting performance of different models, WTI crude oil prices is employed for the sample data set for the experiment. This dataset source is the U.S. Energy Information Administration (EIA; http://www.eia.doe.gov/) and has 8321 observations that include all daily data from 2 January 1986 to 3 January 2019. A test set with 1664 observations is used to evaluate the prediction performance. The daily chart of WTI crude oil prices is illustrated in Figure 6.

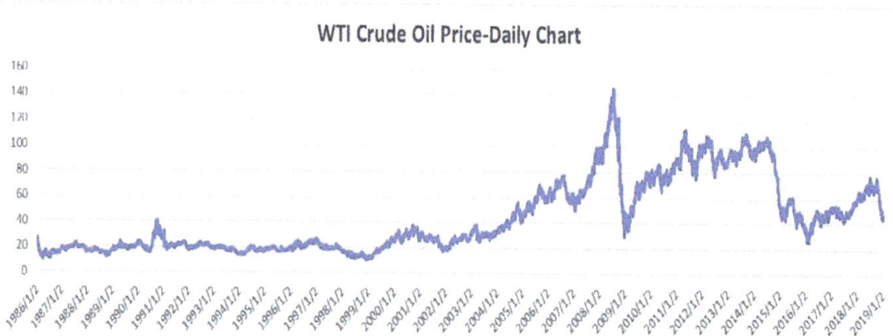

Figure 6. The daily chart of WTI crude oil prices from 2 January 1986 to 3 January 2019.

4.2. Evaluation Metrics and Baselines

Following, we adopt three metrics including the root mean squared error (RMSE), mean absolute percent error (MAPE) and Diebold-Mariano (DM) to evaluate our model.

(1) Root mean squared error (RMSE):

$$\text{RMSE}(y, \hat{y}) = \sqrt{\frac{\sum_{i=1}^{N}(y_i - \hat{y}_i)^2}{N}} \tag{22}$$

RMSE is one of the most common metrics and often used to measure the deviation between observations and prediction in the task of machine learning models. The letter y in the formula above denotes the observations, $\hat{y}$ is the prediction, and N indicate the number of samples.

(2) Mean absolute error (MAE):

$$\text{MAE}(y, \hat{y}) = \frac{\sum_{i=1}^{N}|y_i - \hat{y}_i|}{N} \tag{23}$$

(3) MAE can better reflect the actual situation of the error of prediction. Mean Absolute Error Percentage (MAPE):

$$\text{MAPE}(y, \hat{y}) = \frac{1}{N}\sum_{i=1}^{N}\left|\frac{y_i - \hat{y}_i}{y_i}\right| \tag{24}$$

MAPE measures not only the absolute error between the predicted value and the observations, but also the relative distance. Unlike the previous two metrics, MAPE stands for percentage error, which can help it compare errors between different data sets.

(4) Diebold-Mariano Test (DM test):

$$\text{DM} = \frac{\bar{d}}{\sqrt{\left[\gamma_0 + 2\Sigma_{k=1}^{h-1}\gamma_k\right]/n}} \tag{25}$$

It is used to test the statistical significance of the forecast accuracy of two forecast methods. Variable d is the subtraction of absolute error of the two methods. $\bar{d}$ is the mean of d_i. γ_k is the autocovariance at lag k. DM ~ N(0, 1), if the p-value $> \alpha$, we conclude there is no significant difference between the two forecasts.

An ideal metric is not only reflected in the description of prediction accuracy, but also required to reflect the distribution characteristics of errors. Because of their effectiveness, these metrics are widely used for error measurement in regression or prediction tasks. The definition of each metric is different. When the sample size is large enough, RMSE is more reliable. RMSE is more suitable for Gaussian distribution error measurement than MAE, and MAE has a better performance in measuring uniform error. A single metric is not enough to judge the pros and cons of the model. We need to combine multiple metrics to determine the pros and cons of the model [39].

To verify that the proposed method is advanced, we compare our method with a series of models, most of them mentioned in previous sections. These models include not only the common models such as: naïve forecasts(that the prediction is the same as the last period), ARRIMA, least squares SVR (LSSVR), ANN RNN, LSTM, but also some hybrid models such as: EEMD-ELM, EEMD-LSSVR, EEMD-ANN, EEMD-RNN, EEMD-LSTM and EEMD-SBL-ADD. Both EEMD-LSTM and EMD-SBL-ADD have been proposed in the last two years and represent the current state-of-the-art approaches for crude oil forecasting.

4.3. Experimental Settings

There is one input layer, two hidden layers of GRU and one dense layer in our proposed model. Figure 7 illustrate its Tensorflow computation graph which indicate the network architecture in our method. Each input sample is a matrix of $n \times m$, which is represented by a NumPy array. $'n'$ is the lagging order and $'m'$ is the number of IMFs and residue. By trial and error, we determine set the number of hidden neurons to 32 and MSE as the loss function. The optimizer of training is adaptive moment estimation (Adam) which solve the problems of other algorithm, such as learning rate disappearing, convergence slowly or loss function fluctuating greatly. The learning rate in following experiment is set to 0.01. We adopt the strategy of one day ahead prediction to carry out our tasks. In other words, the prices of the past n days ($p_1, p_2, \ldots p_{n-1}, p_n$) is used to predict the price of the $(n + 1)$th day. Letter n is called the lag order which related to the size of neuron of the GRU. We adopt a strategy of grid search to determine the number of lagging order that is important for time series analysis. By trial and error, the lag order was set to 32.

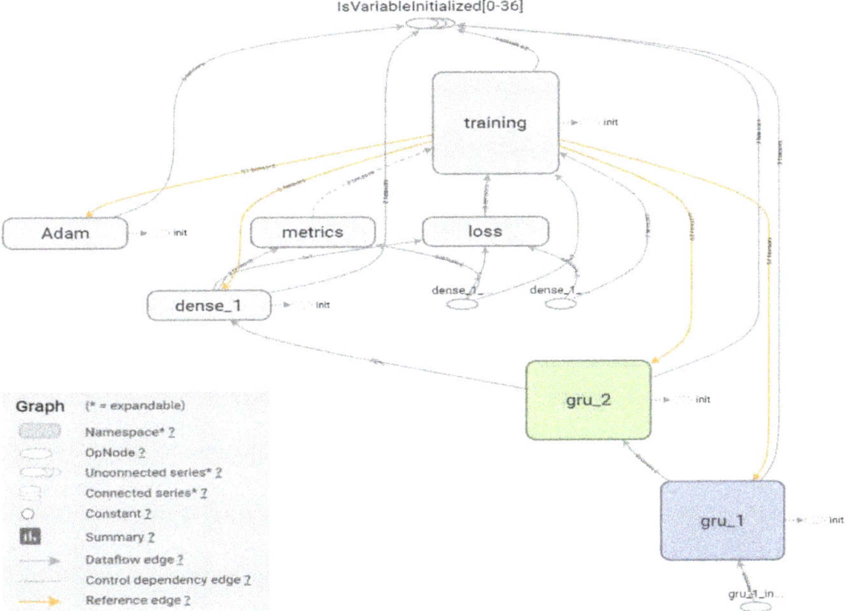

Figure 7. TensorFlow computation graphs of the network structure of proposed model. This figure illustrates the data dependencies and control dependencies. The solid arrows indicate the data dependencies that show the flow of tensors between two ops, while dotted lines indicate the control dependencies.

For comparison purposes, we set the same initialization parameters for several deep learning models that participated in the experiment. For other parameters, since they have little effect on the results, we set the parameters to default.

The methods mentioned in the previous literature will be compared to our introduced approach in the performance of crude oil prices forecasts. These methods include not only the single model, LSSVR, ANN, ARIMA, and LSTM, but also the hybrid method EEMD-LSTM, which is considered the state-out-of- art method in crude oil prices forecasting.

In this study, all experiments are conducted in Python 3.7 via several specialized package, such as TensorFlow-GPU 2.0, Pyeemd 1.4, Keras 2.3 and so on. All experiments are conducted on a PC with a 3.2 GHz CPU, 16 GB RAM and 12 GB CG card.

4.4. Experimental Process, Result and Analysis

4.4.1. Data Decomposition

The original WTI crude oil is decomposed to 13 components which include 12 IMFS and one CEEMDAN residue, where we added the white noise with standard deviation of 0.01 and the number of ensemble size is equal to the size of dataset. Then the IMFs and residue are splinted into the train set and test set. Both them are illustrated in Figure 8, in which the left part of dotted line is train set and the right part is test set.

By trial and error, we determine the parameters of ensemble size of 0.05 and noise strength of 100. It means that the white noise data with standard deviation of 0.05 and quantity of 100 will be added to the original data. Before data decomposition, we initialize the other parameters of the algorithm. Table 1 shows the names and descriptions of the main parameters. Table 2 shows the parameter settings.

Table 1. EEMD and CEEMDAN parameters description.

Parameter	Description
spline_kind	Defines type of spline, which connects extrema
nbsym	Number of extrema used in boundary mirroring
max_imf	IMF number to which decomposition should be performed
ensemble_size	Number of trials or EMD performance with added noise
noise_strength	Standard deviation of the additional noise.

Table 2. Parameters settings.

Method	Nbsym	Max_imf	Trials	Noise_Width/Epsilon
EEMD	2	ALL	100	0.05
CEEMDAN	2	ALL	100	0.05

The time–frequency spectra of IMFs and the residue by CEEMDAN after decomposition are shown in Figure 8. We divide the original data and the decomposed result into a training set and a test set. The former accounts for 80% of the entire dataset, and the latter accounts for the remaining 20%. As shown in Figure 8, the data on the left side of the dotted line constitutes the training set, and the test set is on the right. As one kind of supervised learning algorithm, the input paired to the desired output in both the training and test sets.

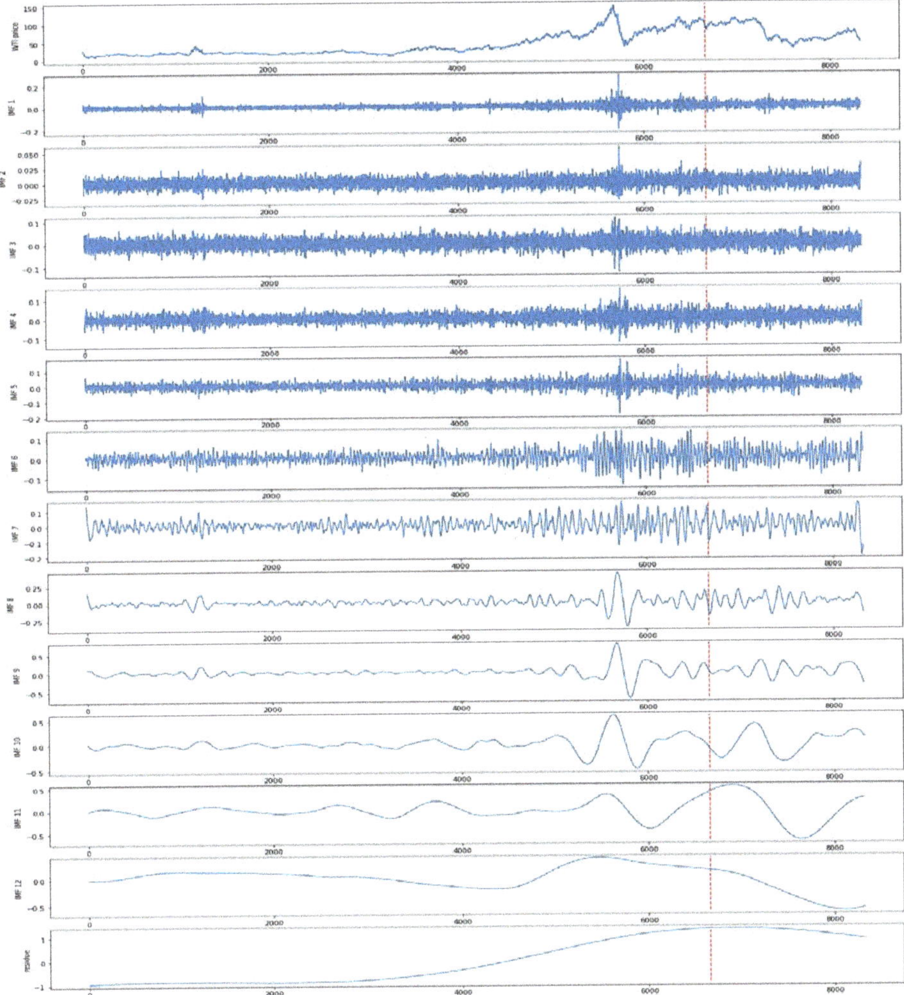

Figure 8. The time–frequency spectrums of IMFs obtained by CEEMDAN. The top sub-picture shows the original WTI price data. The next 12 images represent intrinsic mode functions (IMF), which is listed in the order from the high to the low frequencies. The last component is the residue, which represents the portion of the original data not decomposed and the real trend of the original data.

4.4.2. Training/Learning

After the model is build, we move on to the next step: training/learning. In this process, the training data is continuously feed into the model to incrementally improve the model's predictive performance. Both the loss function and optimizer of adaptive moment estimation (Adam) mentioned earlier are used to evaluate and optimize the model in training to achieve the purpose of model optimization. The workflow of training as shown in Figure 9.

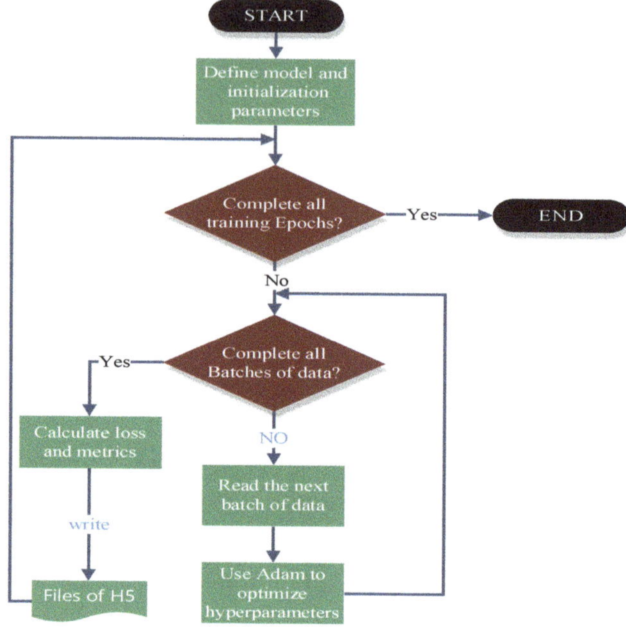

Figure 9. Workflow of training.

The trend of the loss function curve during the iteration of model training from 1st to 100th epoch is shown in the Figure 10.

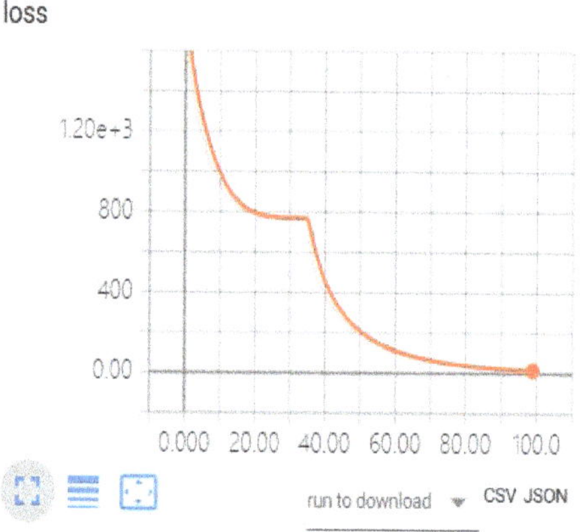

Figure 10. The loss curve graph of model.

The optimization algorithm mentioned above determines the optimization efficiency of the model, while loss represents the distance between the prediction and the observation. From the preceding graph, we observe that the loss curve descends rapidly in the initial epochs of training iteration, which shows that the model is optimized significantly by tuning the hyperparameter. However, between the 25th and 35th epochs of training, the loss curve is flat due to adaptive adjustment of the learning rate. After that, the curve continues to descend rapidly. The curve descending slowly means that loss saturated after 70 epochs of iterations. After 100 training epochs, the loss (MSE) has descended to a very low level. In fact, we conducted 2500 epochs experiment for each model, with 120 samples per batch. Once the training\learning of the hyperparameters (weight) in the model is completed, the model can be used in the tasks of crude oil price forecasts. The experimental results are shown later in the paper.

4.4.3. Predictive Performance of Different Single Model

We propose a hybrid model that includes two components, a decomposition algorithm CEEMDAN and a prediction model Multi-layer GRU. In order to make the method evaluation more accurate, we adopt a combination of independent evaluation and overall evaluation. So, we expand single model comparison and hybrid model comparison between two sets of experiments.

Here, we will evaluate our proposed CEEMD-ML-GRU's effectiveness for improving forecasting accuracy. The compared single models included the naïve forecasts, one classical time series method of ARIMA, two famous machine models of LSSVR and ANN and two popular deep learning model of LSTM and GRU. The samples used in each iteration of the deep learning model are randomly drawn, which makes the results of each prediction different. In order to improve the robustness of the model, experiments with each parameter condition were required to be trained 100 times. In order of performance, the metric at the median position represents the model. The results are shown in Table 3.

Table 3. Predictive performance comparison of single methods.

Metrics	Methods						
	ML-GRU	**GRU**	**LSTM**	**LSSVR**	**ANN**	**ARIMA**	**Naive**
RMSE	**1.2869**	1.4820	1.4818	1.6473	1.5223	2.4861	1.5336
MAE	**1.2424**	1.3817	1.3788	1.5219	1.3649	2.2134	**0.9271**
MAPE	**0.0138**	0.0152	0.0153	0.0168	0.0156	0.0268	0.0152

From Tables 3–5, we can see that:

(1) Among all these models, the multi-layer GRU (ML-GRU) stacking network performed the best on the metrics of RMSE and MAPE. As shown in Table 5, ML-GRU significantly outperformed higher prediction accuracy than other single models. This phenomenon further illustrates that the multi-layer neural network can be used to solve complex issues.

(2) Both GRU and LSTM, designed for long term dependencies of time series show better performance than other traditional machine learning models for crude oil price forecasts. From Table 3, we observe that there is no significant difference between the LSTM and GRU models in the task of crude oil price prediction. Table 4 indicates that GRU has higher efficiency than LSM, because GRU network needs 30% less hyperparameters that need be learned than LSTM, during the training process. It means that GRU neural network use less training parameters comparing to LSTM, and therefore use less resources of computing and storage, execute faster and train faster than LSTM's.

(3) It is interesting that the comprehensive score of the naive forecasts surpasses classical methods such as ARIMA, ANN and LSSVR, and even achieve the best score on MAE. This phenomenon indicates that the complex crude oil price trends are difficult to predict. Misuse of some models, the result is even worse than doing nothing.

(4) While ARIMA performed the worst. The results further confirm that ARIMA, a classic time series analysis model, doesn't perform well at issues of nonlinear and non-stationary time series.

Table 4. Comparison of GRU and LSTM model in the number of parameters.

Method	Input Shape	Output Shape	Total Params
GRU	(32, 13)	(32, 32)	4416
LSTM	(32, 13)	(32, 32)	5888

Table 5. The Diebold–Mariano (DM) test results for single models on WTI crude oil prices.

DM Test	Benchmark Model					
	GRU	**LSTM**	**LSSVR**	**ANN**	**ARIMA**	**Naive**
ML-GRU	−2.1644 (0.02365)	−2.319 (0.02017)	−15.6146 (0.0000)	−3.4532 (0.0000)	−20.198 (0.0000)	−2.1581 (0.02517)

4.4.4. Effect of Selecting Different Hybrid Approaches

On the basis of the previous single model prediction, we continue to evaluate the performance of the hybrid model based on the decomposition method. Table 6 shows the prediction performance of the corresponding hybrid models based on EEMD or CEEMDAN. Table 7 demonstrates that the result of DM test between CEEMDAN-ML-GRU and the other two models.

Table 6. The experimental results in terms of hybrid approaches on WTI crude oil prices forecasting.

Metrics	EEMD			CEEMDAN		
	RMSE	**MAE**	**MAPE**	**RMSE**	**MAE**	**MAPE**
ML-GRU	0.9619	0.9341	0.00987	**0.9276**	**0.9134**	**0.0094**
GRU	0.9912	0.9719	0.0106	0.9334	0.9278	0.0101
LSTM	0.9862	0.965	0.0104	0.9329	0.9261	0.0099
LSSVR	1.1265	1.0847	0.0116	1.1197	1.0903	0.0114
ANN	1.0508	1.0121	0.0106	1.0476	1.0118	0.0105

Table 7. The Diebold–Mariano (DM) test results for hybrid models on WTI crude oil prices.

	Benchmark Model	
DM Test	ML-GRU	EEMD-LSTM EEMD-ML-GRU CEEMDAN-LSTM CEEMDAN- GRU Naive
CEEMDAN-ML-GRU	−6.3309(0.0000)	−1.759(0.0776) −15.6146(0.0000) −3.4532(0.0000) −20.198(0.0000) −11.7541(0.0000)

In this set of experiments, we added a prediction model based on the EEMD decomposition algorithm, and these hybrid models have appeared in the latest literature. Each single model combines CEEMDAN and EEMD respectively, and thus two sets of mixed methods will be obtained. Subsequently, these models were tested in the WTI crude oil price prediction task to derive who is the optimal model. Table 6 demonstrates the predictions of different hybrid models. From the Tables 6 and 7, we can observe that:

(1) Both EEMD and CEEMDAN plus prediction models outperform single model significantly in this prediction task. It indicates that machine learning models with signal processing algorithm based contribute to the better forecasting performance in the nonlinear and non-stationary time series analysis.

(2) The hybrid model based on CEEMDAN has better prediction accuracy than that based on EEMD's. The residual noise from the components decomposed by EEMD, cause a certain extent of reconstruction error, and affect the overall predictive accuracy ultimately.

(3) The ML-GRU with multi-layer architecture still performs better than the GRU with single-layer architecture on the decomposed data. CEEMDAN-ML-GRU, our proposed method has been verified to the best method for the task of crude oil price forecasting.

5. Conclusions

In this study, a hybrid model called CEEMDAN-ML-GRU for crude oil price forecasting is proposed. This model takes full use of the advantages of the signal processing algorithm CEEMDAN and the multi-layer gated recurrent unit networks (ML-GRU). As mentioned in the previous section, the hybrid model uses CEEMDAN to solve the non-stationarity problem of crude oil price data, and generalizes the nonlinear crude oil prices data by a multi-layered GRU neural network. We conduct a large number of experiments to verify the effect of the proposed method in forecasting task by using the WTI price data as sample data. The experimental results show that our proposed method goes beyond other traditional statistical methods, machine intelligent algorithms and other hybrid models, which include the EEMD-LSTM method proposed in 2019.

In addition to crude oil price forecasts, the introduced CEEMDAN-ML-GRU model can also be extended to solve other complex problems in other areas, such as time series forecasts or risk measurements in financial markets. The main purpose of this approach is to improve the accuracy of short-term crude oil price predictions and help decision makers minimize the risks of the crude oil market. However, the proposed method is mainly applied to short-term forecasts, so only daily data is used. If we need to predict long-term price trends, we need to combine this method with economic theory or measurement methods to play a greater role. This is exactly the research plan that we will follow in our future research.

Author Contributions: Conceptualization, H.L.; Formal analysis, Q.S.; Investigation, Q.S.; Methodology, H.L.; Project administration, H.L. and Q.S.; Supervision, Q.S.; Writing—original draft, H.L. All authors have read and agreed to the published version of the manuscript.

Funding: This research was funded by the provincial social science fund of Fujian (No. FJ2018B056), A Study on the Supervision of Taiwan-funded Banks in the Fujian Free Trade Zone, and the National Natural Science Foundation of China (No. 71872046), Environmental Regulation, Enterprise Innovation and Industrial Transformation and Upgrade: Theory and Practice Based on Energy Saving and Emissions Supervision.

Conflicts of Interest: The authors declare no conflict of interest.

References

1. Sadorsky, P. Assessing the impact of oil prices on firms of different sizes: Its tough being in the middle. *Energy Policy* **2008**, *36*, 3854–3861. [CrossRef]

2. Basher, S.A.; Sadorsky, P. Oil price risk and emerging stock markets. *Glob. Financ. J.* **2006**, *17*, 224–251. [CrossRef]

3. Bouri, E.; Kachacha, I.; Roubaud, D. Oil market conditions and sovereign risk in MENA oil exporters and importers. *Energy Policy* **2020**, *137*, 111073. [CrossRef]

4. Wegener, C.; Basse, T.; Kunze, F.; von Mettenheim, H.J. Oil prices and sovereign credit risk of oil producing countries: An empirical investigation. *Quant. Financ.* **2016**, *16*, 1961–1968. [CrossRef]

5. Haushalter, G.D.; Heron, R.A.; Lie, E. Price uncertainty and corporate value. *J. Corp. Finance* **2002**, *8*, 271–286. [CrossRef]

6. Ediger, V.; Serta, A. ARIMA forecasting of primary energy demand by fuel in Turkey. *Energy Policy* **2007**, *35*, 1701–1708. [CrossRef]

7. Krithikaivasan, B.; Zeng, Y.; Deka, K.; Medhi, D. ARCH-Based Traffic Forecasting and Dynamic Bandwidth Provisioning for Periodically Measured Nonstationary Traffic. *IEEE ACM Trans. Netw.* **2007**, *15*, 683–696. [CrossRef]

8. Yu, H.; Shan, G.; Hao, C. Wind Speed Forecasting Based on ARMA-ARCH Model in Wind Farms. *Electricity* **2011**, *22*, 30–34.

9. Zhao, C.L.; Wang, B. Forecasting Crude Oil Price with an Autoregressive Integrated Moving Average (ARIMA) Model. In *Fuzzy Information & Engineering and Operations Research & Management*; Springer: Berlin/Heidelberg, Germany, 2014.

10. Newbold, P. ARIMA Model Building and the Time Series Analysis Approach to Forecasting. *J. Forecast.* **1983**, *2*, 23–35. [CrossRef]

11. Kaiser, T. One-Factor-GARCH Models for German Stocks—Estimation and Forecasting. *Tuebinger Diskussionsbeitraege* **1996**. [CrossRef]

12. Garcia, R.C.; Contreras, J.; Akkeren, M.V.; Garcia, J.B.C. A GARCH forecasting model to predict day-ahead electricity prices. *IEEE Trans. Power Syst.* **2005**, *20*, 867–874. [CrossRef]

13. Mohammadi, H.; Lixian, S. International evidence on crude oil price dynamics: Applications of ARIMA-GARCH models. *Energy Econ.* **2010**, *32*, 1001–1008. [CrossRef]

14. Aamir, M.; Shabri, A. Modelling and Forecasting Monthly Crude Oil Price of Pakistan: A Comparative Study of ARIMA, GARCH and ARIMA Kalman Mode. In *Advances in Industrial and Applied Mathematics*; Salleh, S., Aris, N., Bahar, A., Zainuddin, Z.M., Maan, N., Lee, M.H., Ahmad, T., Yusof, Y.M., Eds.; AIP Publishing LLC: Melville, NY, USA, 2016; Volume 1750.

15. Liu, L. Nonlinear Test and Forecasting of Petroleum Futures Prices Time Series. *Energy Procedia* **2011**, *5*, 754–758.

16. Alquist, R.; Kilian, L.; Vigfusson, R.J. Forecasting the Price of Oil. In *Handbook of Economic Forecasting, Chapter 8*; Elliott, G., Timmermann, A., Eds.; Elsevier: Amsterdam, The Netherlands, 2013; Volume 2, pp. 427–507.

17. Kunze, F.; Spiwoks, M.; Bizer, K.; Windels, T. The usefulness of oil price forecasts—Evidence from survey predictions. *Manag. Decis. Econ. Int. J. Res. Progress Manag. Econ.* **2018**, *39*, 427–446. [CrossRef]

18. Li, T.; Hu, Z.; Jia, Y.; Wu, J.; Zhou, Y. Forecasting Crude Oil Prices Using Ensemble Empirical Mode Decomposition and Sparse Bayesian Learning. *Energies* **2018**, *11*, 1882. [CrossRef]

19. Xie, W.; Yu, L.; Xu, S.; Wang, S. A new method for crude oil price forecasting based on support vector machines. In *International Conference on Computational Science*; Springer: Berlin/Heidelberg, Germany, 2006; pp. 444–451.

20. Fan, L.; Pan, S.; Li, Z.; Li, H. An ICA-based support vector regression scheme for forecasting crude oil prices. *Technol. Forecast. Soc. Chang.* **2016**, *112*, 245–253. [CrossRef]

21. Yu, L.; Wang, S.; Lai, K.K. Forecasting crude oil price with an EMD-based neural network ensemble learning paradigm. *Energy Econ.* **2008**, *30*, 2623–2635. [CrossRef]

22. He, K.; Yu, L.; Lai, K.K. Crude oil price analysis and forecasting using wavelet decomposed ensemble model. *Energy* **2012**, *46*, 564–574. [CrossRef]

23. Hamid, M.H.; Shabri, A. Wavelet Regression Model in Forecasting Crude Oil Price. In *3rd Ism International Statistical Conference 2016*; AbuBakar, S.A., Yunus, R.M., Mohamed, I., Eds.; AIP Publishing LLC: Melville, NY, USA, 2017; Volume 1842.

24. Wu, Y.-X.; Wu, Q.-B.; Zhu, J.-Q. Improved EEMD-based crude oil price forecasting using LSTM networks. *Phys. A Statist. Mech. Appl.* **2019**, *516*, 114–124. [CrossRef]

25. Zhou, Y.; Li, T.; Shi, J.; Qian, Z. A CEEMDAN and XGBOOST-Based Approach to Forecast Crude Oil Prices. *Complexity* **2019**. [CrossRef]

26. Huang, N.E.; Shen, Z.; Long, S.R.; Wu, M.L.C.; Shih, H.H.; Zheng, Q.N.; Yen, N.C.; Tung, C.C.; Liu, H.H. The empirical mode decomposition and the Hilbert spectrum for nonlinear and non-stationary time series analysis. *Proc. R. Soc. a-Math. Phys. Eng Sci.* **1998**, *454*, 903–995. [CrossRef]

27. Wu, Z.; Huang, N. Ensemble Empirical Mode Decomposition: A Noise-Assisted Data Analysis Method. *Adv. Adapt. Data Anal.* **2009**, *1*, 1–41. [CrossRef]

28. Torres, M.E.; Colominas, M.A.; Schlotthauer, G.; Flandrin, P. A complete ensemble empirical mode decomposition with adaptive noise. In Proceedings of the 2011 IEEE International Conference on Acoustics, Speech and Signal Processing (ICASSP), Prague, Czech Republic, 22–27 May 2011; pp. 4144–4147.

29. Zhang, W.; Qu, Z.; Zhang, K.; Mao, W.; Ma, Y.; Fan, X. A combined model based on CEEMDAN and modified flower pollination algorithm for wind speed forecasting. *Energy Convers. Manag.* **2017**, *136*, 439–451. [CrossRef]

30. Cao, J.; Li, Z.; Li, J. Financial time series forecasting model based on CEEMDAN and LSTM. *Phys. A Stat. Mech. Appl.* **2018**. [CrossRef]

31. Duan, J. Financial system modeling using deep neural networks (DNNs) for effective risk assessment and prediction. *J. Franklin Inst.* **2019**, *356*, 4716–4731. [CrossRef]

32. Hosaka, T. Bankruptcy prediction using imaged financial ratios and convolutional neural networks. *Expert Syst. Appl.* **2019**, *117*, 287–299. [CrossRef]

33. Yang, B.; Sun, S.; Li, J.; Lin, X.; Tian, Y. Traffic flow prediction using LSTM with feature enhancement. *Neurocomputing* **2019**, *332*, 320–327. [CrossRef]

34. Fernando, T.; Denman, S.; Sridharan, S.; Fookes, C. Soft + Hardwired attention: An LSTM framework for human trajectory prediction and abnormal event detection. *Neural Netw.* **2018**, *108*, 466–478. [CrossRef]

35. Stafylakis, T.; Khan, M.H.; Tzimiropoulos, G. Pushing the boundaries of audiovisual word recognition using Residual Networks and LSTMs. *Comput. Vis. Image Underst.* **2018**, *176*, 22–32. [CrossRef]

36. Uthayakumar, J.; Metawa, N.; Shankar, K.; Lakshmanaprabu, S.K. Financial crisis prediction model using ant colony optimization. *Int. J. Inf. Manag.* **2020**, *50*, 538–556. [CrossRef]

37. Hochreiter, S.; Schmidhuber, J. Long Short-Term Memory. *Neural Comput.* **1997**, *9*, 1735–1780. [CrossRef] [PubMed]

38. Cho, K.; van Merriënboer, B.; Gulcehre, C.; Bahdanau, D.; Bougares, F.; Schwenk, H.; Bengio, Y. Learning Phrase Representations using RNN Encoder-Decoder for Statistical Machine Translation. *arXiv* **2014**, arXiv:1406.1078. [CrossRef]

39. Chai, T.; Draxler, R.R. Root mean square error (RMSE) or mean absolute error (MAE)? Arguments against avoiding RMSE in the literature. *Geosci. Model Dev.* **2014**, *7*, 1247–1250. [CrossRef]

Article

Optimizing Predictor Variables in Artificial Neural Networks When Forecasting Raw Material Prices for Energy Production

Marta Matyjaszek [1], Gregorio Fidalgo Valverde [2], Alicja Krzemień [3], Krzysztof Wodarski [4] and Pedro Riesgo Fernández [2,*]

[1]   Doctorate Program on Economics and Enterprise, University of Oviedo, Independencia 13, 33004 Oviedo, Spain; marta.matyjaszek@polsl.pl

[2]   School of Mining, Energy and Materials Engineering, University of Oviedo, Independencia 13, 33004 Oviedo, Spain; gfidalgo@uniovi.es

[3]   Department of Risk Assessment and Industrial Safety, Central Mining Institute, Plac Gwarków 1, 40-166 Katowice, Poland; akrzemien@gig.eu

[4]   Faculty of Organization and Management, Silesian University of Technology, Roosevelt 26, 41-800 Zabrze, Poland; krzysztof.wodarski@polsl.pl

*   Correspondence: priesgo@uniovi.es; Tel.: +34-985-10-42-79

Received: 26 March 2020; Accepted: 16 April 2020; Published: 18 April 2020

Abstract: This paper applies a heuristic approach to optimize the predictor variables in artificial neural networks when forecasting raw material prices for energy production (coking coal, natural gas, crude oil and coal) to achieve a better forecast. Two goals are (1) to determine the optimum number of time-delayed terms or past values forming the lagged variables and (2) to improve the forecast accuracy by adding intrinsic signals to the lagged variables. The conclusions clearly are in opposition to the actual scientific literature: when addressing the lagged variable size, the results do not confirm relationships among their size, representativeness and estimation accuracy. It is also possible to verify an important effect of the results on the lagged variable size. Finally, adding the order in the time series of the lagged variables to form the predictor variables improves the forecast accuracy in most cases.

Keywords: raw material; price forecasting; artificial neural network; predictor variable; lagged variable size; rolling window; coking coal; natural gas; crude oil; coal

---

1. Introduction

Artificial neural networks (ANN) have been widely used as accurate forecast aids addressing issues directly or indirectly related with energy or raw materials: energy production [1], raw material inventory levels [2], crude oil prices [3], volatility of stock price indices [4], electricity prices [5], stock prices [6], gold prices [7], copper spot prices [8], off-gases production [9], currency exchange rates [10], etc.

This paper analyzes the forecast of raw material prices for energy production by means of ANN, focusing on the selection of optimum parameters in order to configure the ANN. Design of experiments (DOE) is normally used to select these parameters [11].

DOE can be focused on estimating the number of neurons in hidden layers [12], on forming training and test datasets [13], on eliminating redundant dimensions in the predictor variables trying to achieve compression [14,15], on determining the optimum size of the lagged variables [16], on adding signals to the lagged variables [17], etc.

A heuristic approach will be used to optimize the predictor variables in ANN by means of (1) modifying the lagged variable size and (2) adding intrinsic signals to the lagged variables.

Addressing the lagged variable size, Liu and Su [18] indicated that a larger size allows for increasing the forecast accuracy, while it decreases the representativeness of the subsample heterogeneity. Moreover, although smaller sizes may improve representativeness, they will reduce the estimation accuracy. In the above empirical research, they used lagged variable sizes of 12, 24 and 36 months to test different alternatives. Nevertheless, the lagged variable sizes indicate very little effect on the results.

Tang and Abosedra [19] established that the lagged variable regression results are very sensitive regarding their size, but as there are no proper methods to select an optimum size, arbitrary selections have to be made. Other authors argue that a larger lagged variable size would lead to short-run predictability information being missed, and thus, a shorter size is preferred [20,21].

WEKA from the Machine Learning Group at the University of Waikato (Waikato, New Zealand) [22,23], a well-known open source machine learning software widely used for teaching, research and industrial applications, has a specific time-series analysis environment to forecast models. WEKA's time-series framework uses a machine learning/data mining approach to model time series. It transforms the data by removing the temporal ordering of individual input examples by encoding the time dependency via additional input fields or lagged variables. When using WEKA, it is possible to manipulate and control how lagged variables are created. They are the main mechanism to capture the relationship between current and past values, creating a window over a certain time period. Essentially, the number of lagged variables created determines the size of the window.

Regarding the adding of intrinsic signals to the lagged variables, Tavakoli et al. [24] proposed an input management system based on flexible data by immediately providing a variable definition layer on top of the acquisition layer to feed a data mining module to build modeling functions. Recently, and within the neural networks field, Uykan and Koivo [25] have presented and analyzed a new design for the predictor variables of a radial basis function neural network. In this design, the predictor variables were augmented with a desired output vector, allowing for better/comparable performance when compared with the standard neural network.

Raw material selection, namely, coking coal, natural gas, crude oil and coal, was based on the representativeness and price availability of such materials. In the case of coking coal, prices were obtained from the Colombian Mining Information System as they were publicly disclosed, while for the rest of the raw materials, their prices were obtained from the World Bank Commodity Price Data (The Pink Sheet) under a Creative Commons Attribution 4.0 International License (CC BY 4.0).

The program used to simulate ANNs was NeuralTools 7.5 from Palisade Corporation (Ithaca, NY, USA).

2. Method

This paper will attempt to improve the result of the time-series forecasting of raw material prices for energy production developed with ANN by means of a twofold optimization of the predictor variables (modifying the lagged variable size and adding intrinsic signals to the lagged variables), taking into consideration the previous work of Matyjaszek et al. [26], in which coking coal prices were forecasted by means of autoregressive integrated moving average models (ARIMA) [27,28] and ANN, as well as the transgenic time-series theory.

2.1. Artificial Neural Networks

Two different types of ANNs proposed by Specht [29] will be tested using the best net search function that is available in NeuralTools: generalized regression neural networks (GRNNs), which were used in the past to forecast European thermal coal spot prices among very different applications [30,31], and multilayer feedforward networks (MLFNs), with one or two layers as described in García Nieto et al. [32].

GRNNs are based on nonlinear regression theory and are very closely related to probabilistic neural nets. In GRNNs, a case prediction with a dependent value that is unknown is obtained by means of interpolation from the training cases, with neighboring cases given more weight [33]. The optimal parameters for the interpolation are found during training. The main advantage is not requiring any configuration at all.

Figure 1 presents a GRNN with two independent variables in the input layer and only four training cases, with the pattern layer having four nodes corresponding to these training cases. Each of the nodes will compute its Euclidean distance regarding the presented case. Then, these values pass to the summation layer. The summation layer has two parts: one is the numerator, and the other is the denominator. The numerator contains the sum of multiplying the training output data and the activation function. The denominator is the sum of all the activation functions.

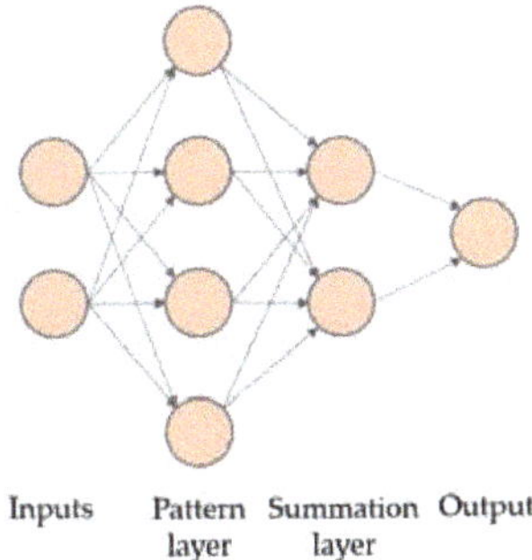

Figure 1. Configuration of a generalized regression neural network (GRNN) with two independent numeric variables and four training cases.

Finally, the output layer has only one neuron that calculates the output by dividing the numerator and the denominator of the summation layer.

On the other hand, MLFNs consist of an input layer, one or even two hidden layers and the output layer. A MLFN is configured by specifying the number nodes in the hidden layers. The net behavior will depend on the number of nodes selected for each hidden layer, the connections weights, the bias terms that are assigned to each node, and the activation/transfer function selected to convert into output the inputs of each node. They are able to approximate complex relationships between the variables.

2.2. Lagged variable Size

One of the issues to be analyzed within this paper is the current discussion about if a larger lagged variable size allows for increasing the forecast accuracy [18] or if a shorter size is preferred based on avoiding to miss short-run predictability information [20,21]. Another issue will be whether lagged variable regression results are very sensitive regarding their size [19] or not [18].

Lagged variables are generated by a number of linear time-delayed input terms or past values, normally in ascending order, such as P(t−n) ... P(t−2), P(t−1), to estimate the output value P(t) [34]. They are also referred to as rolling windows [35].

To undergo a first estimation of the number of time-delayed terms that should form the lagged variables (n), there are several alternatives that can be selected, including the one developed by Ren et al. [36], which uses the seasonal characteristic that appears in the autocorrelation function (ACF) plot, although this value is not always available.

Other common approach is to approximate the value by determining the square root of the amount of data available to undertake the analysis [26]:

$$\sqrt{Total\ n^{\underline{o}}\ of\ data} \tag{1}$$

Another alternative is the one used also by Matyjaszek et al. [26], in which the adequate number of time-delayed terms, k, that should form each input layer is calculated as follows:

$$\text{Total } n^{\underline{o}} \text{ of data} \leq n^2 + 2n + 1,$$ (2)

$$n = 1 + k + 1$$ (3)

2.3. Adding Intrinsic Signals to the Lagged Variables

Up to date and in order to improve the forecast accuracy by adding signals to the lagged variables, research is focused on the extrinsic ones [24,25]. This paper would analyze whether it is feasible to optimize predictor variables by adding intrinsic signals, so that the neural network will have more information available; thus, a better forecast could be made. For this purpose, the order in the time series of each lagged variable would be used, so the ANN could exploit this feature.

This line of thinking is congruent with the work developed by Barabási [37], who states that there is a huge disconnect between network science and deep learning; although ANN are abstractions of natural processes, some of the key neural networks could not be more ignorant about real networks. Main deep learning algorithms treat network features, like degree, as simple variables. Thus, they cannot truly exploit the network effects, which are the essence of these systems, as in networked systems the key information is in the relationships between the connected components (i.e., in the links or edges, which are the direct interactions between nodes), not in the node attributes.

Table 1 presents an example of the first through tenth lagged variables used in a model with five time-delayed input terms: $P(t-k) \ldots P(t-2)$, $P(t-1)$, with $k = 5$, as well as the output to be estimated: $P(t)$.

Table 1. First through tenth lagged variables with five time-delayed input terms, and the output to be estimated (t).

Lagged Variable	$t-5$	$t-4$	$t-3$	$t-2$	$t-1$	t
First	37.93	37.31	34.84	37.87	26.03	38.73
Second	37.31	34.84	37.87	26.03	38.73	40.41
Third	34.84	37.87	26.03	38.73	40.41	38.31
Fourth	37.87	26.03	38.73	40.41	38.31	38.27
Fifth	26.03	38.73	40.41	38.31	38.27	39.33
Sixth	38.73	40.41	38.31	38.27	39.33	39.36
Seventh	40.41	38.31	38.27	39.33	39.36	39.85
Eighth	38.31	38.27	39.33	39.36	39.85	37.30
Ninth	38.27	39.33	39.36	39.85	37.30	38.27
Tenth	39.33	39.36	39.85	37.30	38.27	37.15

Table 2 presents the same first through tenth lagged variables with five time-delayed input terms plus the order in the time series of each lagged variable, as well as the output to be estimated: $P(t)$.

Table 2. First through tenth predictor variables with 5 time-delayed input terms plus the order in the time series of each lagged variable, and the output to be estimated (t).

Neuron Number	Order	$t-5$	$t-4$	$t-3$	$t-2$	$t-1$	t
First neuron	1	37.93	37.31	34.84	37.87	26.03	38.73
Second neuron	2	37.31	34.84	37.87	26.03	38.73	40.41
Third neuron	3	34.84	37.87	26.03	38.73	40.41	38.31
Fourth neuron	4	37.87	26.03	38.73	40.41	38.31	38.27
Fifth neuron	5	26.03	38.73	40.41	38.31	38.27	39.33
Sixth neuron	6	38.73	40.41	38.31	38.27	39.33	39.36
Seventh neuron	7	40.41	38.31	38.27	39.33	39.36	39.85
Eighth neuron	8	38.31	38.27	39.33	39.36	39.85	37.30
Ninth neuron	9	38.27	39.33	39.36	39.85	37.30	38.27
Tenth neuron	10	39.33	39.36	39.85	37.30	38.27	37.15

2.4. Figures of Merit

The experimental results will be evaluated using the two most common figures of merit [38], namely, the root mean squared error (*RMSE*) and the mean absolute error (*MAE*).

The *RMSE* is an excellent general-purpose error measure used for numerical predictions. It amplifies and penalizes large errors and can be expressed as follows:

$$RMSE = \sqrt{\frac{\sum_{t=1}^{n}(A_t - F_t)^2}{n}},\tag{4}$$

where A_t is the actual value, F_t is the forecasted value, and n is the number of forecasted values.

The *MAE* is used to measure how close the predictions are to the outcomes and can be expressed as follows:

$$MAE = \frac{1}{n}\sum_{t=1}^{n}|A_t - F_t|\tag{5}$$

Chai and Draxler [39] proposed the use of a combination of metrics including but not limited to the *RMSE* and the *MAE*. Conversely, Carta et al. [40], when addressing wind resource prediction, proposed using the *MAE*, the *MAPE* and the index of agreement (*IoA*).

In this paper, the standard deviation of absolute error (*STD* of *AE*) was selected to complement these measures as in Lazaridis [38], characterizing the dispersion of the absolute errors.

3. Results

3.1. Coking Coal

The dataset used was the Colombia hard coking coal monthly prices free on board (FOB) for the period from January 1991 to December 2015, as publicly disclosed by the Colombian Mining Information System [41], totaling 300 data points. The dataset is presented in Figure 2.

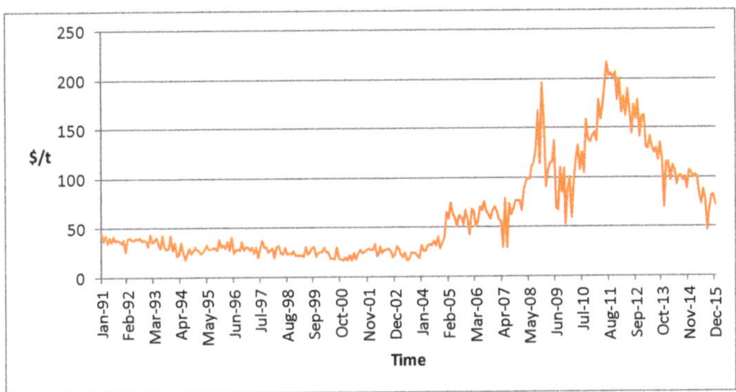

Figure 2. Colombia hard coking coal monthly prices FOB from January 1991 to December 2015 [41].

In first place, GRNNs and MLFNs with 2–6 nodes in the first hidden layer (as the second layer is seldom needed for better prediction accuracy) were tested using the best net search function that is available in NeuralTools.

Table 3 presents the results of this best net search, where the configuration of the lagged variables was made using 19 time-delayed terms, which was the lagged variable size given by the seasonal characteristic that appears in the autocorrelation function (ACF) plot of the transformed time series when representing a consistent genome [26,42]. The results showed that the GRNN improved all the

MLFN models, as stated for most cases by Modaresi et al. [43], so this will be the ANN to be used in this paper.

Table 3. Best net search.

Neural Network Type	*RMS* Error
GRNN	14.41
MLFN 2 Nodes	18.43
MLFN 3 Nodes	24.37
MLFN 4 Nodes	21.91
MLFN 5 Nodes	19.80
MLFN 6 Nodes	19.28

To undergo a first estimation of the number of time-delayed terms that should form the lagged variables, the square root of the amount of data available to undertake the analysis was calculated:

$$\sqrt{Total\ n^{\underline{o}}\ of\ data} = \sqrt{300} = 17.32$$

Using the other alternative previously mentioned, the value obtained is as follows:

$$300 \leq n^2 + n + 1 => n = 16.32,$$

$$k = n - 2 = 14.32$$

Nevertheless, the GRNN will be trained starting with 12 time-delayed input terms and up to 24 time-delayed input terms, a range that includes the previous values of k as well as one and two complete year periods [18].

This allows considering almost any periodical aspect that may be hidden within the time-series values but without drastically reducing the sample size requirements according to Turmon and Fine [44]. The results from the GRNN training are presented in Table 4.

Table 4. Training results for the GRNN model of the Colombian coking coal time series, from 12 to 24 time-delayed input terms (total number of data points is 300).

Time-Delayed Input Terms	Number of Cases	% of Bad Predictions 30% Tolerance	Root Mean Squared Error	Mean Absolute Error	Standard Deviation of Absolute Error
12	288	9.0278%	5.715	3.718	4.340
13	287	9.7561%	6.383	4.295	4.721
14	286	9.7902%	6.352	4.131	4.826
15	285	8.7719%	5.801	3.699	4.469
16	284	9.5070%	6.167	3.996	4.698
17	283	9.1873%	5.678	3.580	4.408
18	282	9.5745%	5.969	3.761	4.636
19	**281**	**8.5409%**	**4.550**	**2.784**	**3.599**
20	280	8.9286%	5.080	3.172	3.968
21	279	8.6022%	4.853	2.949	3.854
22	278	9.7122%	5.352	3.296	4.217
23	277	9.3863%	5.211	3.230	4.089
24	276	9.0580%	4.955	3.040	3.913

The figures in bold correspond to the model that achieves better performance measures.

Based on these measures, the best result was obtained with 19 time-delayed input terms, which was the lagged variable size given by the seasonal characteristic that appears in the autocorrelation function (ACF) plot of the transformed time series when representing a consistent genome [26].

The figures of merit were root mean squared error (*RMSE*) of 4.550, mean absolute error (*MAE*) of 2.784 and standard deviation of absolute error of 3.599. With 30% tolerance, the percentage of bad predictions was 8.5409%.

Then, it was checked if it was feasible to optimize the predictor variables by adding intrinsic signals to the lagged variables so that the ANN would have more information available, and thus, a better forecast could be achieved.

For this purpose, the order of the lagged variables in the time series was considered. Using these predictor variables, results from the training of the GRNN are presented in Table 5.

Table 5. Training results for the GRNN model of the Colombian coking coal time series, from 12 to 24 time-delayed input terms and including the order in the time series of the lagged variables (total number of data points is 300).

Time-Delayed Input Terms	Number of Cases	% of Bad Predictions 30% Tolerance	Root Mean Squared Error	Mean Absolute Error	Standard Deviation of Absolute Error
12	288	7.9861%	5.690	3.816	4.221
13	287	6.9686%	5.599	3.731	4.175
14	286	8.0420%	6.018	4.034	4.466
15	285	8.4211%	6.288	4.246	4.637
16	284	9.5070%	6.112	4.072	4.557
17	283	6.7138%	5.358	3.447	4.102
18	282	5.6738%	5.138	3.311	3.929
19	281	6.4057%	4.914	3.169	3.756
20	280	5.7143%	4.555	2.962	3.460
21	279	3.2258%	3.871	2.545	2.916
22	278	3.2374%	4.575	3.104	3.361
23	277	2.8881%	3.883	2.624	2.863
24	**276**	**2.5362%**	**3.376**	**2.286**	**2.485**

The figures in bold correspond to the model that achieves better performance measures.

The best result was obtained with 24 time-delayed input terms, with a *RMSE* of 3.376, a *MAE* of 2.286 and a standard deviation of absolute error of 2.485. With 30% tolerance, the percentage of bad predictions was 2.5362%. Thus, the order in the time series of the lagged variables clearly improved the model's forecasting performance, as it significantly reduced the *RMSE*, the *MAE*, the standard deviation of absolute error and the percentage of bad predictions.

3.2. Natural Gas

The second raw material for energy production analyzed was natural gas. The dataset used was natural gas prices in Europe for the period from January 1991 to August 2019, totaling 344 values.

Prices were obtained from the World Bank [45] and are presented in Figure 3 in MMBtu, also known as million British thermal units, with 1 MMBtu = 28.263682 m^3 of natural gas at 1 °F.

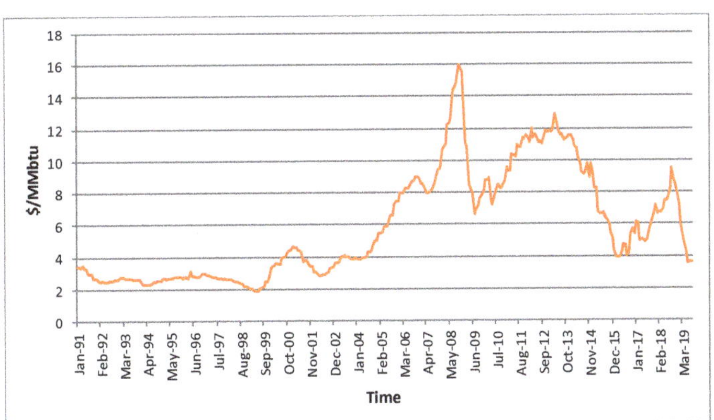

Figure 3. Natural gas prices in Europe for the period from January 1991 to August 2019 [45].

As no seasonal characteristic appears in the autocorrelation function (ACF), in order to estimate the number of time-delayed input terms, the following calculations were made:

$$\sqrt{Total\ n^{\underline{o}}\ of\ data} = \sqrt{344} = 18.55,$$

$$344 \leq n^2 + n + 1 => n = 17.54$$

$$k = n - 2 = 15.54$$

Nevertheless, the GRNN was trained with lagged variables starting with 12 time-delayed input terms and up to 24 time-delayed input terms, using the same interval as in the coking coal case and for the same reasons.

The results from the training of the GRNN are presented in Table 6.

Table 6. Training results for the GRNN model of the European natural gas time series, from 12 to 24 time-delayed input terms (total number of data points is 344).

Time-Delayed Input Terms	Number of Cases	% of Bad Predictions 5% Tolerance	Root Mean Squared Error	Mean Absolute Error	Standard Deviation of Absolute Error
12	**332**	**4.8193%**	**0.06273**	**0.03456**	**0.05235**
13	331	13.2931%	0.11497	0.07372	0.08823
14	330	9.0909%	0.08877	0.04962	0.07361
15	329	11.2462%	0.10689	0.06426	0.08541
16	328	7.3171%	0.07147	0.03534	0.06211
17	327	8.5627%	0.08255	0.04554	0.06885
18	326	7.6687%	0.07654	0.04115	0.06453
19	325	8.9231%	0.08099	0.04326	0.06847
20	324	10.4938%	0.10054	0.05656	0.08312
21	323	9.5975%	0.08318	0.04125	0.07223
22	322	7.1429%	0.07656	0.03844	0.06621
23	321	11.5265%	0.10147	0.05458	0.08553
24	320	11.2500%	0.10862	0.05987	0.09064

The figures in bold correspond to the model that achieves better performance measures.

The best result was obtained with 12 time-delayed input terms, with a *RMSE* of 0.06273, a *MAE* of 0.03456 and a standard deviation of absolute error of 0.05235. With 5% tolerance, the percentage of bad predictions was 4.8193%. A 5% tolerance was used this time, as with a 30% tolerance the percentages of bad predictions were always zero.

Then, the GRNN was trained using the same number of time-delayed input terms but considering the order in the time series of each lagged variable. The results from the training of the GRNN are presented in Table 7. In this case, the best result was obtained with 14 time-delayed input terms, with a *RMSE* of 0.01970, a *MAE* of 0.01032 and a standard deviation of absolute error of 0.01678. With 5% tolerance, the percentage of bad predictions decreased to zero.

Table 7. Training results for the GRNN model of the European natural gas time series, from 12 to 24 time-delayed input terms and including the order in the time series of the lagged variables (total number of data points is 344).

Time-Delayed Input Terms	Number of Cases	% of Bad Predictions 5% Tolerance	Root Mean Squared Error	Mean Absolute Error	Standard Deviation of Absolute Error
12	332	0.9036%	0.05682	0.03368	0.04577
13	331	1.2085%	0.04952	0.03043	0.03907
14	**330**	**0.0000%**	**0.01970**	**0.01032**	**0.01678**
15	329	11.8541%	0.11262	0.07292	0.08583
16	328	2.7439%	0.07011	0.04389	0.05468
17	327	1.2232%	0.04523	0.02703	0.03627
18	326	7.0552%	0.09436	0.05991	0.07291
19	325	4.9231%	0.06879	0.04397	0.05290
20	324	1.8519%	0.06466	0.04194	0.04922
21	323	9.2879%	0.10533	0.06673	0.08150
22	322	7.7640%	0.08618	0.05335	0.06768
23	321	8.4112%	0.09739	0.06137	0.07562
24	320	7.5000%	0.07879	0.04658	0.06354

The figures in bold correspond to the model that achieves better performance measures.

Thus, again, the order in the time series of the lagged variables clearly improved the model's forecasting performance, as it significantly reduced the *RMSE*, the *MAE*, the standard deviation of absolute error and the percentage of bad predictions.

3.3. Crude Oil

The third raw material for energy production analyzed was crude oil. The dataset used was that of Brent crude oil prices for the same period as for the natural gas: January 1991 to August 2019, totaling 344 values. Again, prices were obtained from the World Bank [45] and are presented in Figure 4 in \$/bbl, that is, dollars per barrel, with 1 barrel being approximately 159 liters.

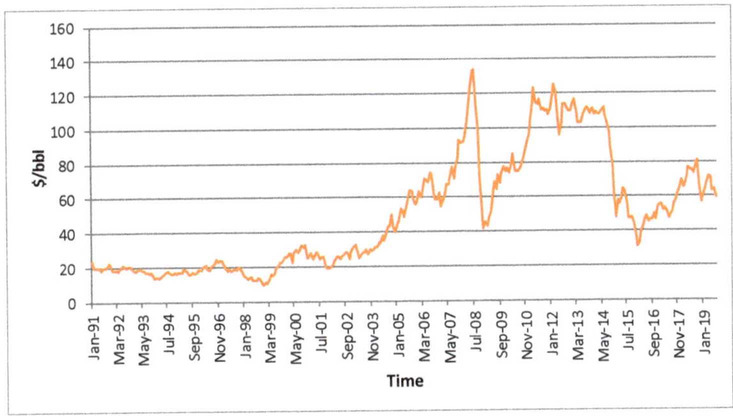

Figure 4. Brent crude oil prices for the period from January 1991 to August 2019 [45].

The number of estimated time-delayed input terms that should be used were the same as in the case of natural gas and crude oil, as the number of data points was the same in all cases (344).

The GRNN was trained again with lagged variables starting with 12 time-delayed input terms and up to 24 time-delayed input terms. The results from the training of the GRNN are presented in Table 8.

Table 8. Training results of the GRNN model for the Brent crude oil time series, from 12 to 24 time-delayed input terms (total number of data points is 344).

Time-Delayed Input Terms	Number of Cases	% of Bad Predictions 5% Tolerance	Root Mean Squared Error	Mean Absolute Error	Standard Deviation of Absolute Error
12	332	29.2169%	1.713	1.1509	1.269
13	331	28.3988%	1.486	0.9643	1.131
14	330	28.7879%	1.601	1.0494	1.209
15	329	29.1793%	1.535	0.9976	1.167
16	328	28.9634%	1.525	0.9911	1.159
17	327	25.6881%	1.324	0.8427	1.021
18	326	29.4479%	1.569	0.9986	1.210
19	325	28.3077%	1.539	0.9680	1.197
20	324	28.0864%	1.563	0.9761	1.221
21	323	28.7926%	1.535	0.9451	1.209
22	322	24.8447%	1.316	0.7827	1.058
23	**321**	**23.0530%**	**1.177**	**0.6645**	**0.971**
24	320	26.8750%	1.420	0.8398	1.145

The figures in bold correspond to the model that achieves better performance measures.

The best result was obtained with 23 time-delayed input terms, with a *RMSE* of 1.177, a *MAE* of 0.6645 and a standard deviation of absolute error of 0.971. With 5% tolerance, the percentage of bad predictions was 23.0530%.

Then, the GRNN was trained using the same number of time-delayed input terms but considering the order in the time series of each lagged variable. The results from the training of the GRNN are presented in Table 9.

Table 9. Training results of the GRNN model for the Brent crude oil time series, from 12 to 24 time-delayed input terms including the order in the time series of the lagged variables (total number of data points is 344).

Time-Delayed Input Terms	Number of Cases	% of Bad Predictions 5% Tolerance	Root Mean Squared Error	Mean Absolute Error	Standard Deviation of Absolute Error
12	332	0.0000%	0.0000	0.0000	0.0000
13	331	0.0000%	0.1708	0.1147	0.1265
14	**330**	**0.0000%**	**0.0281**	**0.0178**	**0.0218**
15	329	4.8632%	0.5863	0.3764	0.4496
16	328	17.0732%	0.9034	0.5742	0.6974
17	327	4.2813%	0.4508	0.2649	0.3648
18	326	12.5767%	0.7637	0.4729	0.5996
19	325	5.8462%	0.7225	0.4589	0.5581
20	324	0.6173%	0.2725	0.1653	0.2166
21	323	5.5728%	0.6292	0.3951	0.4896
22	322	1.8634%	0.4352	0.2698	0.3414
23	321	13.0841%	0.8028	0.5109	0.6193
24	320	0.6250%	0.3644	0.2332	0.2801

The figures in bold correspond to the model that achieves better performance measures.

The best result was obtained with 14 time-delayed input terms, with a *RMSE* of 0.0281, a *MAE* of 0.0178 and a standard deviation of absolute error of 0.0218. With 5% tolerance, the percentage of bad predictions decreased again to zero.

With 12 time-delayed input terms, it is clear that the ANN was able to learn the exact configuration of the time series, but only in this case.

Thus, the order in the time series of the lagged variables clearly improved the model's forecasting performance, as it significantly reduced the *RMSE*, the *MAE*, the standard deviation of absolute error and the percentage of bad predictions.

3.4. Coal

The fourth and last raw material for energy production analyzed was coal. The dataset used was that of Australian coal prices for the same period as crude oil and natural gas: January 1991 to August 2019. Prices were also obtained from the World Bank [45] and are presented in Figure 5 in $/t.

Figure 5. Australian coal prices for the period January from 1991 to August 2019 [45].

The number of estimated time-delayed input terms that should be used were the same as in the case of natural gas and crude oil, as the number of data points was the same in all cases (344).

The GRNN was trained again with lagged variables starting with 12 time-delayed input terms and up to 24 time-delayed input terms, as in the case of European natural gas and Brent crude oil. The results from the training of the GRNN are presented in Table 10.

Table 10. Training results of the GRNN model for the Australian coal time series, from 12 to 24 time-delayed input terms (total number of data points is 344).

Time-delayed Input Terms	Number of Cases	% of Bad Predictions 5% Tolerance	Root Mean Squared Error	Mean Absolute Error	Standard Deviation of Absolute Error
12	332	17.1687%	1.4800	0.9232	1.1568
13	331	20.8459%	1.6634	1.0467	1.2928
14	330	16.6667%	1.4359	0.8757	1.1379
15	329	21.5805%	1.6421	1.0063	1.2976
16	328	21.6463%	1.6679	1.0163	1.3225
17	327	19.2661%	1.5443	0.9240	1.2373
18	326	16.8712%	1.4328	0.8257	1.1710
19	**325**	**15.6923%**	**1.3280**	**0.7397**	**1.1029**
20	324	24.3827%	1.7703	1.0805	1.4023
21	323	23.8390%	1.7435	1.0541	1.3888
22	322	21.1180%	1.6711	0.9920	1.3448
23	321	20.2492%	1.6219	0.9430	1.3196
24	320	18.4375%	1.5812	0.9129	1.2910

The figures in bold correspond to the model that achieves better performance measures.

The best result was obtained with 19 time-delayed input terms, with a *RMSE* of 1.3280, a *MAE* of 0.7397 and a standard deviation of absolute error of 1.1029. With 5% tolerance, the percentage of bad predictions was 15.6923%.

Then, the GRNN was trained using the same number of time-delayed input terms but considering the order in the time series of each lagged variable. The results from the training of the GRNN are presented in Table 11.

Table 11. Training results for the GRNN models of the Australian coal time series, from 12 to 24 time-delayed input terms and including the order in the time series of the lagged variables (total number of data points is 344).

Time-Delayed Input Terms	Number of Cases	% of Bad Predictions 5% Tolerance	Root Mean Squared Error	Mean Absolute Error	Standard Deviation of Absolute Error
12	332	18.6747%	1.5567	1.0548	1.1448
13	331	18.4290%	1.5274	1.0027	1.1521
14	330	17.8788%	1.4749	0.9614	1.1185
15	329	17.6292%	1.3702	0.8624	1.0647
16	328	23.1707%	1.8393	1.1910	1.4017
17	327	22.6300%	1.7316	1.1060	1.3324
18	326	21.4724%	1.6358	1.0383	1.2641
19	325	20.9231%	1.5696	0.9819	1.2245
20	324	20.3704%	1.4961	0.9180	1.1813
21	323	18.5759%	1.4415	0.8794	1.1423
22	**322**	**16.7702%**	**1.3678**	**0.8185**	**1.0959**
23	321	23.0530%	1.9438	1.1856	1.5404
24	320	22.5000%	1.8997	1.1497	1.5123

The figures in bold correspond to the model that achieves better performance measures.

The best result was obtained with 22 time-delayed input terms, with a *RMSE* of 1.3678, a *MAE* of 0.8185 and a standard deviation of absolute error of 1.0959. With 5% tolerance, the percentage of bad predictions was 16.7702%.

In this case, adding the order in the time series of each lagged variable only improved the standard deviation of absolute error.

Nevertheless, if the *RMSE* and *MAE* were compared with the training results obtained without adding the order in the time series of each lagged variable, although slightly higher, they were very similar to the best results, and better than the rest of the training results.

Thus, the differences between both options were almost negligible.

4. Discussion and Conclusions

This paper applied a heuristic approach to optimize the predictor variables in artificial neural networks when forecasting raw material prices for energy production to achieve a better forecast.

Two goals are (1) to determine the optimum number of time-delayed terms or past values forming the lagged variables and (2) to optimize predictor variables by adding intrinsic signals to the lagged variables.

The experimental results were evaluated using the two most common figures of merit, the root mean squared error (*RMSE*) and the mean absolute error (*MAE*), as well as the standard deviation of absolute error, as the scientific literature proposes the use of a combination of metrics including *RMSE* and *MAE*, but not being limited to them.

Results demonstrated, first, that in opposition to scientific literature when addressing lagged variable size, a larger size did not allow for increasing the forecast accuracy, and that smaller sizes did not reduce the estimation accuracy. Moreover, the lagged variable regression results were very sensitive regarding their size.

In the three raw materials with the same number of cases (natural gas, crude oil and coal), the best results were obtained with rolling window sizes of 12, 23 and 19, respectively. Furthermore, it was possible to verify an important effect of the lagged variable size on the results, with differences that in some cases were larger than 20%.

Thus, and in opposition again to scientific literature indicating that there are no proper methods to select an optimum size so arbitrary selections have to be made, it is recommendable to address this question by trial and error method, although the approximate size can be estimated in order to select the complete year's period range to which this value belongs, e.g., 12–24 months or 24–36 months. This will allow considering any periodical aspect that may be hidden within the time series values, but without drastically reducing or increasing the sample size requirements for neural networks.

Second, in three of the four raw materials analyzed (coking coal, natural gas and crude oil), it was possible to improve the forecast accuracy by adding the order in the time series of the lagged variables to form the predictor variables. The best results were achieved with rolling window sizes of 24, 14 and 14, respectively.

In the case of the Australian coal, this process only improved the standard deviation of absolute error. Nevertheless, if the *RMSE* and *MAE* were compared with the training results obtained without adding the order, although slightly higher, they were very similar to the best results and better than the rest of the training results without adding the order.

As the differences between both options were almost negligible, it is possible to recommend adding the order in the time series of each lagged variable to the predictor variable in all cases.

Third, only with the Brent crude oil with 12 time-delayed input terms and considering the order in the time series of the lagged variables, the ANN was able to learn or deduct the exact configuration of the time series. This is completely congruent with the fact that there is a huge disconnect between network science and deep learning, as the key information is in the relationships between the connected components, not in the node attributes.

Concluding, the findings presented in this paper have an immediate practical application addressing the forecast of time series by means of ANN that consider lagged variables, without being restricted to the studied case of raw material prices for energy production.

Any forecast may be optimized just by adding an intrinsic signal to the predictor variable consisting of the order in the time series of each lagged variable. By doing this way, the ANN will be able to exploit this feature, something that will not happen otherwise. In most of the cases, figures of merit may improve (may be reduced) up to a 20%, with the consequent benefit for decision-makers regarding savings, efficiency/benefit gains and/or lower risk.

Regarding the size of the lagged variable, a selection should be made about the period that will be analyzed in order to undergo a trial and error process. This selection should follow the procedure shown in this paper or other ones that may be found in the scientific literature.

Further research should address different issues such as the use of more intrinsic signals. Regarding this issue, authors have made interesting preliminary approaches by considering the transgenic time series theory that allows eliminating anomalous phenomena from the time series. Augmenting the lagged variables within this anomalous period with a '1' and the rest with a '0', or vice versa, it was possible to improve a priori the figures of merit.

Another area of interesting future research will be to develop a procedure to determine accurately the number of time-delayed input terms that should be used when considering the order in the time series of the lagged variables. While the seasonal characteristic that appears in the autocorrelation function (ACF) plot is valid before augmenting the lagged variables, later this figure is no longer valid, so a new approach should be addressed. Nevertheless, nothing is yet developed addressing the time series with an ACF plot that does not allow one to extract a seasonal characteristic. Again, the transgenic time series theory could be of help in these cases.

Finally, it should be addressed by future research why in the case of Australian coal, or in similar cases, it was not possible to improve the figures of merit by adding to the predictor variable the order in the time series of each lagged variable.

Author Contributions: Conceptualization, M.M. and P.R.F.; methodology, A.K. and P.R.F.; software, G.F.V.; validation, K.W.; formal analysis, G.F.V.; investigation, A.K., P.R.F. and M.M.; resources, G.F.V.; data curation, G.F.V.; writing—original draft preparation, M.M. and P.R.F.; writing—review and editing, A.K.; visualization, P.R.F.; supervision, K.W. and A.K.; and project administration, P.R.F. and K.W. All authors have read and agree to the published version of the manuscript.

Funding: This research received no external funding.

Acknowledgments: The doctoral research fellowship from the Technological Research Foundation "Luis Fernández Velasco" in Mineral and Mining Economics at the School of Mining, Energy and Material Engineering from the University of Oviedo (Spain) for the first author is highly appreciated.

Conflicts of Interest: The authors declare no conflict of interest.

References

1. Bermejo, J.F.; Fernández, J.F.G.; Polo, F.O.; Márquez, A.C. A review of the use of artificial neural network models for energy and reliability prediction. A study of the solar PV, hydraulic and wind energy sources. *Appl. Sci.* **2019**, *9*, 1844. [CrossRef]

2. Ali, S.M.; Paul, S.K.; Azeem, A.; Ahsan, K. Forecasting of optimum raw material inventory level using artificial neural network. *Int. J. Oper. Quant. Manag.* **2011**, *17*, 333–348.

3. Gabralla, L.; Abraham, A. Computational modeling of crude oil price forecasting: a review of two decades of research. *Int. J. Comput. Inf. Syst. Ind. Manag. Appl.* **2013**, *5*, 729–740.

4. Hyup Roh, T. Forecasting the volatility of stock price index. *Expert Syst. Appl.* **2007**, *33*, 916–922. [CrossRef]

5. Panapakidis, I.P.; Dagoumas, A.S. Day-ahead electricity price forecasting via the application of artificial neural network based models. *Appl. Energy* **2016**, *172*, 132–151. [CrossRef]

6. Hadavandi, E.; Shavandi, H.; Ghanbari, A. Integration of genetic fuzzy systems and artificial neural networks for stock price forecasting. *Knowledge-Based Syst.* **2010**, *23*, 800–808. [CrossRef]

7. Mombeini, H.; Yazdani-chamzini, A. Modeling Gold Price via Artificial Neural Network. *J. Econ. Bus. Manag.* **2015**, *3*, 3–7. [CrossRef]

8. Sánchez Lasheras, F.; de Cos Juez, F.J.; Suárez Sánchez, A.; Krzemień, A.; Riesgo Fernández, P. Forecasting the COMEX copper spot price by means of neural networks and ARIMA models. *Resour. Policy* **2015**, *45*, 37–43. [CrossRef]

9. Colla, V.; Matino, I.; Dettori, S.; Cateni, S.; Matino, R. Reservoir Computing Approaches Applied to Energy Management in Industry. In *Engineering Applications of Neural Networks, Proceedings of the 20th International Conference EANN, Xersonisos, Crete, Greece, 24–26 May 2019*; Macintyre, J., Iliadis, L., Maglogiannis, I., Jayne, C., Eds.; Springer: Cham, Switzerland, 2019; pp. 66–79.

10. Baffour, A.A.; Feng, J.C.; Taylorb, E.K. A hybrid artificial neural network-GJR modeling approach to forecasting currency exchange rate volatility. *Neurocomputing* **2019**, *365*, 285–301. [CrossRef]

11. Sánchez Lasheras, F.; Vilán Vilán, J.A.; García Nieto, P.J.; del Coz Díaz, J.J. The use of design of experiments to improve a neural network model in order to predict the thickness of the chromium layer in a hard chromium plating process. *Math. Comput. Model.* **2010**, *52*, 1169–1176. [CrossRef]

12. Bo, L.; Cheng, C. First-Order Sensitivity Analysis for Hidden Neuron Selection in Layer-Wise Training of Networks. *Neural Process. Lett.* **2018**, *48*, 1105–1121.

13. Pontes, F.J.; Amorim, G.F.; Balestrassi, P.P.; Paiva, A.P.; Ferreira, J.R. Design of experiments and focused grid search for neural network parameter optimization. *Neurocomputing* **2016**, *186*, 22–34. [CrossRef]

14. Li, F.; Zurada, J.M.; Liu, Y.; Wu, W. Input Layer Regularization of Multilayer Feedforward Neural Networks. *IEEE Access* **2017**, *5*, 10979–10985. [CrossRef]

15. Li, F.; Zurada, J.M.; Wu, W. Smooth group L1/2 regularization for input layer of feedforward neural networks. *Neurocomputing* **2018**, *314*, 109–119. [CrossRef]

16. Xu, X. Price dynamics in corn cash and futures markets: cointegration, causality, and forecasting through a rolling window approach. *Financ. Mark. Portf. Manag.* **2019**, *33*, 155–181. [CrossRef]

17. Velázquez Medina, S.; Carta, J.A.; Portero Ajenjo, U. Performance sensitivity of a wind farm power curve model to different signals of the input layer of ANNs: Case studies in the Canary Islands. *Complexity* **2019**, *2019*, 2869149.

18. Liu, G.D.; Su, C.W. The dynamic causality between gold and silver prices in China market: A rolling window bootstrap approach. *Financ. Res. Lett.* **2019**, *28*, 101–106. [CrossRef]

19. Tang, C.F.; Abosedra, S. Tourism and growth in Lebanon: new evidence from bootstrap simulation and rolling causality approaches. *Empir. Econ.* **2016**, *50*, 679–696. [CrossRef]

20. Timmermann, A. Elusive return predictability. *Int. J. Forecast.* **2008**, *24*, 1–18. [CrossRef]

21. Tang, C.F.; Chua, S.Y. The savings-growth nexus for the Malaysian economy: A view through rolling sub-samples. *Appl. Econ.* **2012**, *44*, 4173–4185. [CrossRef]

22. Frank, E.; Hall, M.A.; Witten, I.H. *The WEKA Workbench. Online Appendix for "Data Mining: Practical Machine Learning Tools and Techniques"*, 4th ed.; Morgan Kaufmann: Cambridge, MA, USA, 2016.

23. Hall, M.; Frank, E.; Holmes, G.; Pfahringer, B.; Reutemann, P.; Witten, I.H. The WEKA Data Mining Software: An Update. *ACM SIGKDD Explor. Newsl.* **2009**, *11*(1), 10–18. [CrossRef]

24. Tavakoli, S.; Mousavi, A.; Komashie, A. Flexible data input layer architecture (FDILA) for quick-response decision making tools in volatile manufacturing systems. *IEEE Int. Conf. Commun.* **2008**, *1*, 5515–5520.

25. Uykan, Z.; Koivo, H.N. Analysis of Augmented-Input-Layer RBFNN. *IEEE Trans. Neural Netw.* **2005**, *16*, 364–369. [CrossRef]

26. Matyjaszek, M.; Riesgo Fernández, P.; Krzemień, A.; Wodarski, K.; Fidalgo Valverde, G. Forecasting coking coal prices by means of ARIMA models and neural networks, considering the transgenic time series theory. *Resour. Policy* **2019**, *61*, 283–292. [CrossRef]

27. Hyndman, R.; Khandakar, Y. Automatic time series forecasting: The forecast package for R. *J. Stat. Softw.* **2008**, *27*, 1–22. [CrossRef]

28. Ong, C.S.; Huang, J.J.; Tzeng, G.H. Model identification of ARIMA family using genetic algorithms. *Appl. Math. Comput.* **2005**, *164*, 885–912. [CrossRef]

29. Specht, D.F. A general regression neural network. *IEEE Trans. Neural Netw.* **1991**, *2*, 568–576. [CrossRef]

30. Krzemień, A.; Riesgo Fernández, P.; Suárez Sánchez, A.; Sánchez Lasheras, F. Forecasting European thermal coal spot prices. *J. Sustain. Min.* **2015**, *14*, 203–210. [CrossRef]

31. Krzemień, A. Dinamic fire risk prevention strategy in underground coal gasification processes by means of artificial neural networks. *Arch. Min. Sci.* **2019**, *64*, 3–19.

32. García Nieto, P.J.; Sánchez Lasheras, F.; García-Gonzalo, E.; de Cos Juez, F.J. PM10 concentration forecasting in the metropolitan area of Oviedo (Northern Spain) using models based on SVM, MLP, VARMA and ARIMA: A case study. *Sci. Total Environ.* **2018**, *621*, 753–761. [CrossRef]

33. Panella, M.; Barcellona, F.; D'Ecclesia, R.L. Forecasting Energy Commodity Prices Using Neural Networks. *Adv. Decis. Sci.* **2012**, *2012*, 1–26. [CrossRef]

34. Li, Z.; Best, M. Optimization of the Input Layer Structure for Feed-Forward Narx Neural Networks. *Int. J. Electr. Comput. Energ. Electron. Commun. Eng.* **2015**, *9*, 673–678.

35. Morantz, B.H.; Whalen, T.; Zhang, G.P. A weighted window approach to neural network time series forecasting. In *Neural Networks in Business Forecasting*; Zhang, G.P., Ed.; IRM Press: Hershey, PA, USA, 2004.

36. Ren, Y.; Suganthan, P.N.; Srikanth, N.; Amaratunga, G. Random vector functional link network for short-term electricity load demand forecasting. *Inf. Sci.* **2016**, *367*, 1078–1093. [CrossRef]
37. Barabási, A.-L. *Network Science*, 1st ed.; Cambridge University Press: Cambridge, UK, 2016.
38. Lazaridis, A.-G. Prosody Modelling using Machine Learning Techniques for Neutral and Emotional Speech Synthesis. Ph.D. Thesis, University of Patras, Patras, Greece, 2011.
39. Chai, T.; Draxler, R.R. Root mean square error (RMSE) or mean absolute error (MAE)? -Arguments against avoiding RMSE in the literature. *Geosci. Model Dev.* **2014**, *7*, 1247–1250. [CrossRef]
40. Carta, J.A.; Cabrera, P.; Matías, J.M.; Castellano, F. Comparison of feature selection methods using ANNs in MCP-wind speed methods. A case study. *Appl. Energy* **2015**, *158*, 490–507. [CrossRef]
41. System of Colombian Mining Information. Available online: http://www1.upme.gov.co/simco/Paginas/default.aspx (accessed on 11 May 2017).
42. Riesgo García, M.V.; Krzemień, A.; Manzanedo del Campo, M.Á.; Escanciano García-Miranda, C.; Sánchez Lasheras, F. Rare earth elements price forecasting by means of transgenic time series developed with ARIMA models. *Resour. Policy* **2018**, *59*, 95–102. [CrossRef]
43. Modaresi, F.; Araghinejad, S.; Ebrahimi, K. A Comparative Assessment of Artificial Neural Network, Generalized Regression Neural Network, Least-Square Support Vector Regression, and K-Nearest Neighbor Regression for Monthly Streamflow Forecasting in Linear and Nonlinear Conditions. *Water Resour. Manag.* **2018**, *32*, 243–258. [CrossRef]
44. Turmon, M.J.; Fine, T.L. Sample Size Requirements for Feedforward Neural Networks. In Proceedings of the Advances in Neural Information Processing Systems 7, Denver, CO, USA, 28 November 28–1 December 1994; pp. 1–18.
45. The World Bank. Available online: http://pubdocs.worldbank.org/en/561011486076393416/CMO-Historical-Data-Monthly.xlsx (accessed on 24 September 2019).

Article

Prediction of Health-Related Leave Days among Workers in the Energy Sector by Means of Genetic Algorithms

Aroa González Fuentes [1], Nélida M. Busto Serrano [2,*], Fernando Sánchez Lasheras [3,*], Gregorio Fidalgo Valverde [4] and Ana Suárez Sánchez [4]

[1] School of Mining, Energy and Materials Engineering of Oviedo, University of Oviedo, 33007 Oviedo, Spain; UO212081@uniovi.es
[2] Labor and Social Security Inspectorate, Ministry of Labor and Social Economy, 33007 Oviedo, Spain
[3] Mathematics Department, Faculty of Sciences, University of Oviedo, 33007 Oviedo, Spain
[4] Department of Business Administration, School of Mining, Energy and Materials Engineering of Oviedo, University of Oviedo, 33007 Oviedo, Spain; gfidalgo@uniovi.es (G.F.V.); suarezana@uniovi.es (A.S.S.)
* Correspondence: nelida.busto@mitramiss.es (N.M.B.S.); sanchezfernando@uniovi.es (F.S.L.); Tel.: +34-985-103-376 (F.S.L.)

Received: 3 April 2020; Accepted: 12 May 2020; Published: 14 May 2020

Abstract: In this research, a model is proposed for predicting the number of days absent from work due to sick or health-related leave among workers in the industry sector, according to ergonomic, social and work-related factors. It employs selected microdata from the Sixth European Working Conditions Survey (EWCS) and combines a genetic algorithm with Multivariate Adaptive Regression Splines (MARS). The most relevant explanatory variables identified by the model can be included in the following categories: ergonomics, psychosocial factors, working conditions and personal data and physiological characteristics. These categories are interrelated, and it is difficult to establish boundaries between them. Any managing program has to act on factors that affect the employees' general health status, process design, workplace environment, ergonomics and psychosocial working context, among others, to achieve success. This has an extensive field of application in the energy sector.

Keywords: sick leave; absenteeism; energy sector; genetic algorithms (GA); multivariate adaptive regression splines (MARS)

---

1. Introduction

Over the past few years, the main concerns of industry, especially in developed countries, have been to improve the workers' productivity, occupational health and safety in the workplace, physical and mental well-being, and job satisfaction. Through the application of ergonomics, it has been shown that these issues have improved, so an effective implementation of ergonomics in the workplace can achieve a balance between the characteristics of the worker and the demands of the task, in addition to improving workplace design and introducing appropriate management programs. Companies that belong to the energy sector have also been working in this direction, showing some distinctive features that have been identified and studied here.

When the industry does not get involved in the abovementioned issues, it can affect the lives of workers. This results in a risk of deterioration in health and causes absenteeism. Currently, one of the major concerns is sick leave.

There are several factors that affect sick leaves. In this introduction, we first revise the studies that explain the behavior of such factors in the general working environment. Afterwards, we focus on the few works that are specifically oriented to the singularities of the energy sector.

According to the EUROSTAT [1], some 1194 M€ have been spent on sickness and health benefits in the European Union. This number was equivalent to 8.0% of gross domestic product (GDP). The average expenditure per inhabitant was 2338 Euros. In recent years, the spending on sick leave benefits has increased to 12.4%, with Norway being the country with the highest spending (32.4%) compared to Portugal, the country with the lowest (5.8%). For these reasons, the problem of absenteeism is of great interest to healthcare professionals, employers and economists.

According to the latest research published by the European Commission [2], women have higher rates of sick leave than men. There are multiple reasons for this. They have more precarious work and work contracts often linked to low income. Moreover, women often seek medical help for less-serious illnesses and are more frequently diagnosed with mental-health-related illnesses. Another explanation is connected to the burden of housework and childcare.

The aforementioned study [2] also mentions that sick leave increases with age; elderly people take longer-term leave compared to young people, who generally take short-term leave. This is due to the fact that health worsens with age, and working conditions have deteriorated since the economic recession. A clear correlation between occupational, socioeconomic status and absence due to illness is highlighted: the more physically demanding the occupation is and the lower the socioeconomic status, the higher the absenteeism.

Regarding the energy sector, different factors have been studied that may lead to sick leave. A study carried out in the petroleum industry [3] corroborates that women are more likely to take sick leave than men. This may be mainly due to three factors: there is sick leave due to pregnancy; they tend to have more temporary contracts; and they suffer more psychological problems. Another risk factor is smoking; workers who are smokers or former smokers have a higher risk of taking sick leave than non-smokers, those who consume alcohol and even workers exposed to chemical products. Other good predictors of absenteeism due to illness are abnormal sleep and job dissatisfaction. Some physical activity is recommended for workers.

Another specific factor that has been studied in the energy sector is how the different shiftwork patterns can cause sick leave. The employees of a power plant prefer 12-hour shifts rather than eight-hour shifts, since they enjoy longer breaks and an improved social and domestic life. This measure also improves mood, health status and both the quality and quantity of sleep. The only drawback is that it can pose a potential safety risk for the employee when performing highly demanding tasks at the end of the shift, since concentration decreases [4]. A study conducted among workers of a nuclear power plant [5] confirms all these findings on the relation between working in shifts and domestic, social life and well-being of the worker.

Another relevant factor in the energy sector is the type of work carried out. It has been proven that people who do manual work, also called blue-collar (production) workers, are more likely to remain on long-term leave, and this would be of longer duration than in those people who perform skilled jobs, also known as white-collar workers (office workers and managers). The latter are exposed to a greater risk for short-term pain, mainly musculoskeletal disorders (MSD), but this could be avoided by correcting posture [6].

The design of the job is another very important factor to take into account in this sector. A study of workers in a thermal power plant [7] detected deficiencies both in its facilities and its resources. This type of industry is complex and usually has more problems than other industries. Apart from health problems due to ergonomic factors, it has been found that production work in combination with bad environmental conditions (excessive temperatures in summer) and very noisy and dusty environments are factors that tend to worsen the health of workers and increase the possibility of their going on sick leave.

In summary, talking about the energy sector, it seems that absenteeism rates have different behaviors attending to factors like gender, working organization circumstances, such as shifts or psychosocial demands, workplace environment and other factors related with workstation design. The way these factors combined affect sick leave remains unexplored.

The use of machine-learning techniques to predict occupational health and safety outcomes in different fields is not new, whether focused on work-related accidents [8], fire risk [9,10], MSD [11–14] or visual disorders [15,16]. Some of these works focus on specific sectors, such as mining or the health industry.

Nevertheless, to date, no researches have studied how the combination of factors such as age, gender, well-being, domestic and social life, as well as psychosocial factors, can influence the proneness to sick leave among workers in the energy sector. As far as it is known by the authors, most of the previous research in this field that make use of machine-learning techniques employed just only, for example, support vector machines [8,16], artificial neural networks [9,11], Multivariate Adaptive Regression Splines (MARS) [10,14] or k-nearest neighbors [12]. All these research studies have shown the utility of machine-learning techniques in this area for both regression and classification. However, until today, there have been few works [13,15] that combine more than one machine-learning methodology in order to improve their performance.

In this research, a hybrid methodology that combines MARS and genetic algorithm is proposed for predicting the number of days absent from work due to sick or health-related leave, among workers in the industry sector, according to ergonomic, social and work-related factors reported in the Sixth European Working Conditions Survey (EWCS).

2. Materials and Methods

2.1. Dataset

This research work employs selected microdata from the Sixth European Working Conditions Survey (EWCS), which was conducted in 2015 by the European Foundation for the Improvement of Living and Working Conditions, Eurofound [17]. The EWCS is generally conducted every five years, providing an overview of working conditions of the European population. A random representative sample of "persons in employment" (i.e., employees and the self-employed) is surveyed through a questionnaire administered face-to-face. The Eurofound datasets are stored and promoted online by the UK Data Service [18]. Upon request, the data are available free of charge, provided they will be used for non-commercial purposes.

Almost 44,000 workers in 35 countries were interviewed through the sixth wave of the EWCS. The validity of the questionnaire was guaranteed by a questionnaire-development group composed of experts and representatives of the European Commission and different international organizations [19]. This sixth edition codified more than 370 variables that included physical and psychosocial risk factors, working time, place of work, work-pace determinants, employee participation, job security, social relations, personal conditions, etc. The whole list of variables included in the sixth wave of the survey can be found in the source questionnaire, available online at the website of Eurofound [20]: https://www.eurofound.europa.eu/sites/default/files/page/field_ef_documents/6th_ewcs_2015_final_source_master_questionnaire.pdf.

The size of the initial dataset was first reduced by only selecting workers from energy-related sectors. The final sample consisted of 420 workers (333 men and 87 women), aged between 17 and 71 years (average 44; see Figure 1) from the following NACE Revision 1 sections: mining of coal and lignite; extraction of peat; extraction of crude petroleum and gas; mining of uranium and thorium ores; manufacture of coke, refined petroleum products and nuclear fuel; electricity, gas, steam and hot water supply. The distribution of the subjects by country is shown in Table 1. Table 2 presents their level of studies, according to the International Standard Classification of Education (ISCED). Table 3 shows the distribution of the sample of workers according to their household's total monthly income. The average leave time was of 5.9 days, with a standard deviation of 16.1 days. Only two workers have a leave longer than 100 days. Please also note that leaves over 10 days represent only 18.33% of the total.

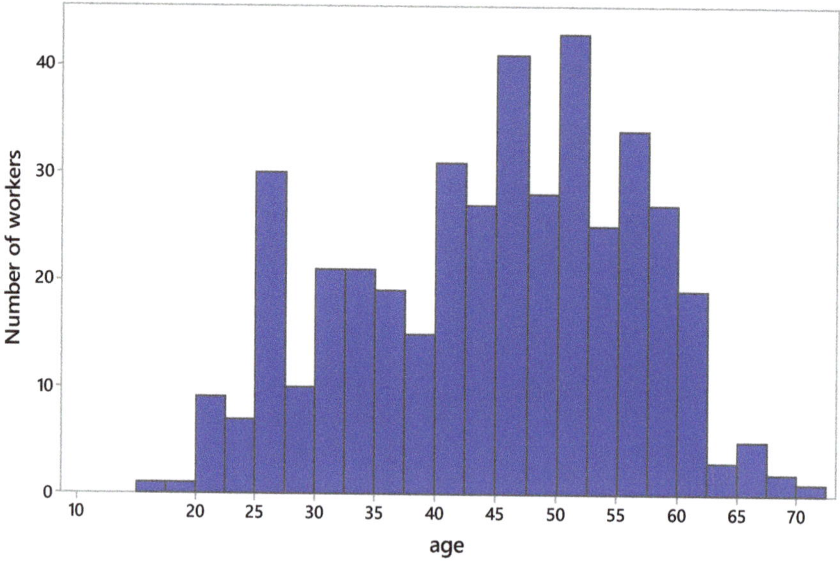

Figure 1. Age distribution of the workers.

Table 1. Distribution of the sample of workers by country.

Country	Number of Workers	%
Norway	35	8.3%
Albania	26	6.2%
Slovenia	25	6.0%
Serbia	25	6.0%
Spain	22	5.2%
Belgium	18	4.3%
United Kingdom	17	4.0%
Croatia	17	4.0%
Montenegro	17	4.0%
Czech Republic	15	3.6%
Poland	14	3.3%
Bulgaria	13	3.1%
Germany	13	3.1%
France	12	2.9%
Romania	12	2.9%
Denmark	11	2.6%
Lithuania	11	2.6%
Austria	11	2.6%
Estonia	10	2.4%
Other countries *	76	18.1%
Total	**420**	

* Other countries: FYROM, Luxembourg, Slovakia, Hungary, Netherlands, Sweden, Switzerland, Greece, Turkey, Italy, Finland, Cyprus, Portugal, Malta.

A second-dimensional reduction was carried out by decreasing the number of variables through expert criteria. Only the 59 most relevant independent variables were preselected to initially feed the model developed and to try to explain the output variable. Some of the variables were designed as Likert scales, some of them were binary and a few were continuous (numerical). Please note that it would have been possible to perform this reduction also by means of either genetic algorithms or other methodologies like decision trees or PCA, but in our understanding, it was less time-consuming to use expert criteria. Please note that this is a good way in order to avoid finding spurious relationships.

The output variable, y15_Q82, records the answers provided by the sample of workers to the following question: "In the past 12 months, how many days absent from work due to sick leave or health-related leave?" It is a numerical variable, ranging from 0 to 360, and synthetizes the duration of the sick leave taken by the workers.

Table 2. Distribution of the sample of workers according to the level of studies (ISCED).

Level of Studies (ISCED)	Number of Workers	%
Early childhood education	1	0.2%
Primary education	1	0.2%
Lower secondary education	39	9.3%
Upper secondary education	185	44.0%
Post-secondary non-tertiary education	23	5.5%
Short-cycle tertiary education	54	12.9%
Bachelor or equivalent	52	12.4%
Master or equivalent	63	15.0%
Doctorate or equivalent	2	0.5%
Total	**420**	

Table 3. Distribution of the sample of workers according to their household's total monthly income.

Is Your Household Able to Make Ends Meet?	Number of Workers	%
Very easily	55	13.1%
Easily	101	24.0%
Fairly easily	121	28.8%
With some difficulty	104	24.8%
With difficulty	30	7.1%
With great difficulty	9	2.1%
Total	**420**	

2.2. Multivariate Linear Regression

Let us consider a set of $k + 1$ quantitative variables with y as the dependent variable and $x_1, x_2, \ldots, x_k$ as independent variables. The multivariate linear regression method consists of creating a lineal model that predicts y, using variables $x_1, x_2, \ldots, x_k$. It can be expressed as follows [21]:

$$y = \beta_0 + \beta_1 x_1 + \ldots + \beta_k x_k \tag{1}$$

The parameters' estimation is performed by means of the ordinary least squares approach [22] by means of the following:

$$min_{\beta \in \mathbb{R}^{k+1}} \| y - X\beta \|^2 = min_{\beta \in \mathbb{R}^{k+1}} \sum_{i=1}^{n} \left(y_i - \beta_0 - \sum_{j=1}^{k} \beta_j x_{ij} \right)^2 \tag{2}$$

where $\| \cdot \|$ denotes the Frobenius norm.

2.3. Support Vector Machine for Regression

Let us consider again a set of $k + 1$ quantitative variables with y as the dependent variable and $x_1, x_2, \ldots, x_k$ as independent variables, where each i element constitutes a row vector. Let $\Phi : \chi \to \mathcal{F}$ be the function that corresponds to each row vector with a point of the characteristics space $\mathcal{F}$. Let us define a function as follows [23]:

$$f(x) = \, < \omega, \Phi(x) > + b \tag{3}$$

The problem to solve is as follows:

$$\min\frac{1}{2}\|\omega\|^2 + C\sum_{i=1}^{n}\left(\xi_i + \xi_i^*\right) \tag{4}$$

Constrains:

$$\begin{array}{ll} y_i - <\omega,\phi(x_i)> \le \varepsilon + \xi_i & i = 1,\ldots,n \\ <\omega,\phi(x_i)> -y_i > \le \varepsilon + \xi_i^* & i = 1,\ldots,n \\ \xi_i \ge 0, \ \xi_i^* \ge 0 & i = 1,\ldots,n \end{array} \tag{5}$$

The problem complexity depends on the dimension of the row vectors [24]. The solution of this problem gives as a result the model of support-vector machines for regression.

2.4. Genetic Algorithms

The process of learning by trial and error can be considered as being similar to the natural evolution process. The development of genetic algorithms (GA) started with the works of Holland [25]. GA is a kind of evolutionary algorithm that is based on the evolution of a certain set of solutions trying to either maximize of minimize the result of an objective function. GA is a bioinspired methodology that mimics the procedure of natural selection. The interest in GA methodology in optimization is because they are a global and robust method for finding solutions that do not require any a priori knowledge about the problem.

GA make use of the following three basic operators [16]:

- Crossover;
- Mutation;
- Elitism.

The crossover operator takes two different individuals of the population and creates a new one, mixing the two. The mutation operator performs random changes in those individuals, created with the help of the crossover operator. Mutation makes it possible to introduce new strings in the next generation, giving the ability to search beyond the scope of the initial population. Another interesting mechanism is elitism, which makes a certain number of individuals with a good performance according to the result of the objective function survive and pass to the next generation, without any change.

2.5. Multivariate Adaptive Regression Splines

MARS is a well-known parametric methodology that builds a non-linear model based on hinge functions. It is expressed by the following equation [26]:

$$\hat{y}_j = \beta_0 + \sum_{i=1}^{k}\beta_i \cdot B(x_i), \tag{6}$$

where $\hat{y}_j$ represents each one of the outputs forecasted values per each y_j, β_i are the model parameters and B are the model basis functions. The basis functions are defined as follows:

$$\begin{aligned} B^- &= \begin{cases} (t-x)^q & if \ x < t \\ 0 & otherwhise \end{cases} \\ B^+ &= \begin{cases} (t-x)^q & if \ x \ge t \\ 0 & otherwhise \end{cases} \end{aligned} \tag{7}$$

where q is a natural number that represents the power function.

When a MARS model is created, there are three different well-known methods that take part in the model in order to assess the importance of the variables. They are the following:

- nsubsets: this criterion indicates the number of model subsets that make use of the variable. The larger the number of subsets that include the variable, the more important they will be considered.
- gcv: this criterion calculates the generalized cross-validation (GCV) of the variables, and, taking into account the results, those variables that contribute most to increasing the GCV value are considered the most important.
- rss: this criterion can be considered equivalent to gcv, but making use of the residual sum of squares (RSS) expression.

The GCV expression is as follows [23]:

$$GCV(M) = \frac{\frac{1}{n} \cdot \sum_{i=1}^{n} \left(y_j - \hat{y}_j \right)}{\left(\frac{1 - C(M)}{n} \right)^2},$$ (8)

where $C(M)$ is the complexity penalty function that increases with the number of basis functions in the model and which is defined with this equation, where M is the number of basis functions. In the case of the present research, the maximum interaction degree allowed was nine.

The equation for RSS is as follows [27]:

$$RSS = GCV(M) \cdot \frac{3 \cdot M}{n^3}$$ (9)

2.6. The Proposed Algorithm

The proposed algorithm works as is explained here. First, it is initialized with a random population. Each member of the random population represents a subset of all the available variables that will be employed for the forecast of the number of days off for each worker. It is a string, as in the following example: 1100011 … 0101 with a total of 59 digits, one per variable, where 1 means that the variable is present and 0 that is missing.

In order to know how each of the variables subsets performs, they are employed for training a MARS model, using 80% of the available individuals, while the other 20% are employed for the model validation. This process is repeated 1000 times for each of the variables subset and the average R^2 value obtained is used as the result of the objective function. Following the usual methodology of genetic algorithms, the best individuals of the population are selected and crossed.

In the present research, a mutation rate of 10% was allowed, and a 5% of elitism, which means that the 5% of the best individuals of a generation are included in the next one. In the case of the present research, a fine-tuning was performed, testing mutation rates from 0.5% to 15% in steps of 0.5%. The R^2 values obtained did not find statistically significant differences from 0.5% to 10%, while in higher mutation values, the R^2 decreased. Therefore, results with 10% probability mutation rates are presented. The crossover methodology employed is known as *single point crossover*, in which both parental chromosomes are split in only one point randomly selected. Each generation of the genetic algorithm population is formed by 1000 individuals; this means that there are 1000 different variables subsets. The results shown were obtained after 100 iterations, which means that 100,000 variables subsets were examined. This is a small number if compared with the more than 5.7×10^{17} possible variables subsets that can be obtained for a problem like this with 59 independent variables. Finally, it can also be highlighted that the performance of the algorithm would be improved if those workers whose leave durations are over certain threshold value (i.e., 10 or 100 days) were considered as outliers and removed. The flowchart for the proposed algorithm is shown in Figure 2.

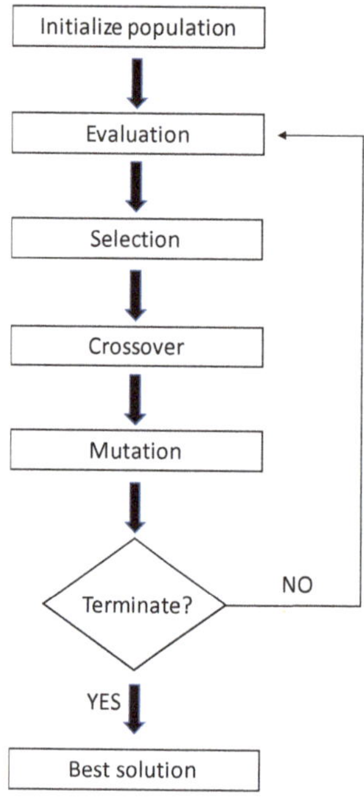

Figure 2. Flowchart of the proposed algorithm.

3. Results

The average value of the R^2 for the 1000 models trained for the variables that were finally selected in each case was 74.26%. The RMSE average value was 27.51. The average difference in days in absolute value of forecasted and real number of days off was 10.22. If workers are divided in those with leaves of 10 or less days, 85% of the total, and those with leaves of more than 10 days, the RMSE values obtained are quite different. In the case of those with leaves of 10 or less days, the RMSE value was of 5.56 and 71.49 for those with leaves of more than 10 days. In the case of average difference in days in absolute value of forecasted and real number of days off, it was 4.28 for those with leaves of 10 or less days and 48.81 for those whose leaves are over 10 days. It means that prediction for those leaves of 10 or less days are much more accurate. Figure 3 shows the histogram of the number of leave days for all the workers. It can be observed that most of the leaves are of 10 or less days. The results described in this paragraph are in line with what is said in Section 2.3 about how an outlier's removal would improve the algorithm performance.

In order to assess the performance of the proposed algorithm, it was compared with two benchmark methodologies: linear regression [28] and support vector machines for regression [29]. In both algorithms, 1000 models with different training and validation datasets were tested. For the linear regression, the average R^2 value was of 26.72% with an RMSE of 74.21, while in the case of support vector machine for regression the average R^2 value was of 67.32% with a RMSE of 29.01. Table 4 shows a comparison of the performance of the proposed algorithm with the two benchmark methodologies referred before, linear regression and support vector machines.

Figure 4 shows the results of one of the models created with these variables. In such a model, the forecasted and real number of days off for all the workers randomly included in the validation dataset can be observed. The values in the horizontal axis are the workers' identificators. Please note that the largest differences of forecasted and real values can be found for workers with numbers from 81 to 83 that would be considered as spurious. In this case, the difference in days of forecasted and real number of days off was on average −4.255, with a median of 2.53 and a standard deviation of 27.34. In absolute values, the mean was 10.642 days. As after 100 iterations all the models give similar results in terms of R^2, Table 5 shows the list of variables selected by this model as the most relevant when predicting absenteeism among workers in the energy sector, using the three importance criteria (nsubsets, gcv and rss) referred to in Section 2.2.

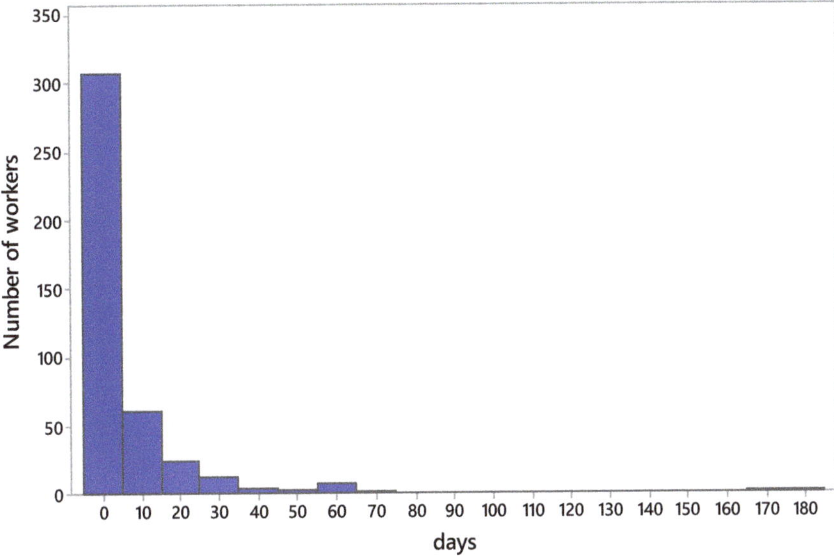

Figure 3. Histogram of the number of leave days of all those individuals that took part in the present research.

Table 4. Comparison of the performance of the proposed algorithm with two benchmark methodologies, linear regression (LR) and support vector machines (SVM).

Performance Metric	Proposed Algorithm	LR	SVM
R^2	74.26%	26.72%	67.32%
RMSE			
all	27.51	74.21	29.01
10 or less leave days	5.56	72.92	13.91
more than 10 leave days	71.49	74.47	68.75
Average absolute difference of days			
all	10.22	49.60	17.26
10 or less leave days	4.28	47.62	10.06
more than 10 leave days	48.81	61.51	60.42

Table 5. A list of variables selected by the model as the most relevant to predict absenteeism among workers in the energy sector.

Variable	Nsubsets [1]	gcv [2]	rss [3]	Description
y15_Q48a	30	100	100	Short repetitive tasks of less than 1 min
y15_Q88	30	100	100	Satisfied with working conditions
y15_Q100	30	100	100	Income
y15_Q20	29	90.3	91.0	Restructuring or reorganization at the workplace
y15_Q53d	29	90.3	91.0	Monotonous tasks
y15_Q78f	29	90.3	91.0	Headaches, eyestrain
y15_Q95f	27	81.7	82.5	Outside work: taking a training or education

Table 5. *Cont.*

Variable	Nsubsets [1]	gcv [2]	rss [3]	Description
y15_Q75	24	62.6	65.2	General health status
y15_Q48b	19	45.6	49.4	Short repetitive tasks of less than 10 min
y15_Q30a	18	41.7	46.0	Tiring or painful positions
y15_Q30h	17	38.5	43.2	Being in situations that are emotionally disturbing
y15_Q76	17	38.5	43.2	Long-lasting illness
y15_Q29e	15	34.8	39.3	Smoke, fumes, powder, dust
y15_Q30d	14	30.9	36.2	Sitting
y15_Q27_lt	13	27.9	33.6	Other paid job
y15_Q74	12	24.8	31.0	Work affects health
y15_Q29b	9	15.9	23.9	Noise
y15_Q24	8	13.9	22.0	Work hours/week
y15_Q29d	5	13.4	18.2	Low temperatures
y15_Q2b	4	11.2	15.9	Age

[1] nsubsets: number of model subsets that make use of the variable (see Section 2.2). [2] gcv: generalized cross-validation of the variables (see Section 2.2). [3] rss: residual sum of squares (see Section 2.2).

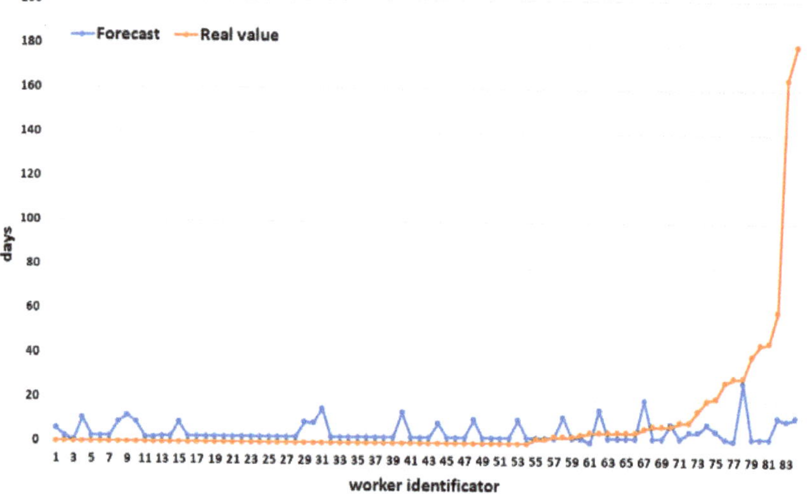

Figure 4. Results of one of the models created: forecasted and real number of days off for all the workers randomly included in the validation dataset.

4. Discussion

The results obtained show that it is possible to make use of machine-learning methodologies such as MARS and GA in order to predict the number of days of health-related leave taken by workers in the energy sector, taking into account a certain number of variables linked to personal and work-related factors for each individual. The results obtained are not surprising, as GA and MARS have proven to be valid in similar scenarios [23] and in other problems linked to the energy field [30,31]. The worst forecasts were obtained for the longest leaves, as they were few and present a behavior outside the normal range. In future research, the MARS model could be substituted by other regression models such as neural networks or support vector machines for regression, and even by replacing GA with other evolutive methods like particle swarm optimization or differential evolution.

As other studies have shown [13,32], there are several factors that have an influence on absenteeism in the workplace. It should be pointed out that, in this case study, sick leave due to common illness and sick leave due to occupational illness or occupational accidents were not analyzed separately. This is the reason behind the multiple kinds of factors that become part of the model, and implies that some difficulties may appear during the discussion of the results [33]; in any case, it is necessary to consider both, in order to understand the causes behind absenteeism.

The model was built with twenty items that can be classified into four categories:

- Ergonomics;
- Psychosocial factors;
- Working conditions;
- Personal data and physiological characteristics.

It must be pointed out that several of these items could belong to different categories simultaneously. Therefore, they are included in those ones that better explain their impact as a cause of absenteeism. Moreover, there are many interrelated circumstances between each category, to the extent that it is difficult to find studies that only deal with one of them. In fact, there are several research studies that cover issues in relation to ergonomics while speaking about working conditions, workers' personal characteristics and organizational contexts at the same time [32].

In absolute terms, working conditions have the greatest impact on sick leave among workers, since nine out of twenty items of the model developed fall into this category. However, as mentioned before, several of these items also affect other categories, such as ergonomics and psychosocial factors.

A discussion on the relationship between absenteeism and the items in each category is presented next.

4.1. Ergonomics

There are five items in the model that show the impact of ergonomics on sick leave among workers. This is in consonance with other researchers' conclusions [33–36] that maintain that poor ergonomics in the workplace and prevalence of musculoskeletal disorders (MSD) are linked and could therefore mean an increase in sick leave. These five items are as follows:

- Doing short repetitive tasks of less than 1 min.
- Doing monotonous tasks.
- Doing short repetitive tasks of less than 10 min.
- Suffering tiring or painful positions.
- Remaining seated for a long time.

It is remarkable that the model seems to suggest a contradictory idea in relation to repetitive tasking, in that it considers that short repetitive tasks of less than ten minutes have a negative impact on sick leave among workers, whereas short repetitive tasks of less than one minute can reduce absenteeism. Concerning ergonomics, the shorter the task, the more damaging it could be, so at first glance, this would seem to be an error.

On the other hand, there could be several explanations for this curious result. For instance, it is difficult to find a job that requires doing the same repetitive tasks lasting less than one minute for the entire working day. However, it is more feasible to find jobs that include the same repetitive task lasting less than ten minutes for the entire working day. Thus, multitasking would preferably be linked to the first of these cases, and multitasking is a valued characteristic of good ergonomics. This would be a proper explanation for the behavior of these items in the model. In any case, this sets a starting point for future research.

Other items that are included in the model and classified into the ergonomics category, such as monotonous tasks, painful positions and sitting, are ergonomic factors traditionally considered during risks assessments due to their negative impact on MSD prevalence. As a first conclusion, the model has proven that the beliefs as to how ergonomic investments have a positive impact on MSD prevalence and occupational absenteeism are a step in the right direction. This study therefore supports other researchers' conclusions that are applicable to industrial environments [37] and to the energy sector in particular [38].

4.2. Psychosocial Factors

Only one item falls into this category: that of being in situations that are emotionally disturbing. However, there are several studies that point out that psychosocially demanding working environments have a deeply negative impact on absenteeism [39,40]. Nevertheless, it should be noted that this does not mean that psychosocial factors are less relevant than others. In fact, there is a close interrelation between them and the rest. For instance, as other studies have shown [41], ergonomic interventions could be counterproductive, unless they attend to psychosocial factors.

4.3. Working Conditions

Working conditions is the category that includes the largest number of items from the model. However, as previously stated, that does not necessarily mean it is the category with the strongest influence on absenteeism. The nine items from the model classified in this category are as follows:

- Being satisfied with working conditions;
- Income;
- Restructuring or reorganization at the workplace;
- Working environment: smoke, fumes, powder and dust;
- Another paid job;
- Work affects health;
- Noise;
- Work hours/week;
- Working environment: low temperatures.

First of all, it must be said that "being satisfied with working conditions" could be included in the category of psychosocial factors; after all, this item depends both on how working conditions are designed and how workers perceive them. In any case, this aspect has already been discussed in the previous section. In fact, the item included on working conditions highlights the interaction between categories and the importance of psychosocial factors on the control of absenteeism.

Working conditions cover a great spectrum of factors that can become causes of occupational absenteeism. Indeed, workers' general health status, MSD prevalence and other diseases are closely related to the working environment. Therefore, in terms of the energy sector in particular, each company has to understand and act on several working conditions, to develop health management, as other works have already shown [42].

A conclusion that can be obtained by analyzing the factors that appear in this category is, as has been the case with others, the existence of a link with the category concerning personal conditions. For example, it has been proven that a low income of the worker increases the probability of his/her taking sick leave. One wonders if, in fact, this circumstance reveals the effects of socioeconomic status on occupational absenteeism [43].

Other working condition factors included in the model show the effects of workplace environment on sick leave: noise, air pollution, extreme temperatures, etc. This was to be expected, because it points in the direction of occupational diseases, in line with several previous studies [44,45].

The only remarkable item that could be seen as a contradiction is that workers with more than one job seem to be less vulnerable to sick leave. Several explanations can be suggested. For instance, having more than one job could be linked with multitasking and, therefore, lower prevalence of MSD. Another possible reason could be that this kind of worker is more likely to belong to a precarious social stratum where health damages are sometimes underreported. In any case, this could be the subject of a future line of research.

4.4. Personal Data and Physiological Characteristics

There are five possible causes of absenteeism included in this category:

- Having headaches and eyestrain;
- Outside work: receiving training or education;
- General health status;
- Suffering a long-lasting illness;
- Age.

This category joins together several factors related to lifestyle that are too difficult to identify and analyze properly, especially when the studied variable includes both common and occupational diseases as a cause of absenteeism. In fact, there are many studies that deal with this subject without achieving unanimity [46–48].

It seems to be expected that general health status [49] and, as a result, other factors that can alter it, like age, must affect the prevalence of several illnesses that end up causing sick leave. There are other studies that go further and that try to analyze gender differences in this matter [14]. In this category, however, everything is vaguely interrelated, so there are many questions to answer in future research.

Apart from the item that refers to activities outside work, every factor included in this category is a cause or an effect of the workers' general health status. This appraisal has many implications that must be kept in mind when planning any move designed to reduce absenteeism among a company's workforce.

5. Conclusions

It is possible to make use of machine-learning methodologies, such as MARS and GA, in order to create models able to predict the number of days of health-related leave among workers in the energy sector. Absenteeism can be monitored and predicted by using a model that employs several items included in the following categories:

- Ergonomics;
- Psychosocial factors;
- Working conditions;
- Personal data and physiological characteristics.

These categories are all interrelated, and it is difficult to establish boundaries between them, but as a positive consequence of this, acting on one of them to reduce absenteeism in a company could have a great impact on the others.

Any management program has to act on factors that affect the employees' general health status, process design, workplace environment, ergonomics and psychosocial working context, among others, if it is going to be successful. This has an extensive field of application in the energy sector, where most of the activities are undertaken in an industrialized context.

Author Contributions: Conceptualization, F.S.L.; formal analysis, F.S.L.; investigation, A.G.F., N.M.B.S. and A.S.S.; methodology, F.S.L. and A.S.S.; project administration, A.S.S.; resources, A.G.F.; validation, G.F.V.; writing—original draft, A.G.F. and N.M.B.S.; writing—review and editing, G.F.V. and A.S.S. All authors have read and agree to the published version of the manuscript.

Funding: This research received no external funding.

Acknowledgments: The authors wish to acknowledge the European Foundation for the Improvement of Living and Working Conditions (Eurofound), as well as the UK Data Service, for providing us with the results of the database used in this work.

Conflicts of Interest: The authors declare no conflict of interest.

References

1. Social Protection Statistics—Sickness and Health Care Benefits. Available online: https://ec.europa.eu/eurostat/statistics-explained/index.php?title=Social_protection_statistics_-_sickness_and_health_care_benefits#Relative_importance_of_expenditure_on_sickness_and_healthcare_benefits (accessed on 3 January 2020).

2. European Comission. Sick Pay and Sickness Benefit Scheme in the European Union. Background Report for the Social Protection Committee's. In-Depth Review on Sickness Benefits. Available online: https://ec.europa.eu/social/BlobServlet?docId=16969&langId=en (accessed on 3 January 2020).

3. Oenning, N.; Carvalho, F.M.; Lima, V.M.C. Risk factors for absenteeism due to sick leave in the petroleum industry. *Rev. Saúde Pública* **2014**, *48*, 103–112. [CrossRef] [PubMed]

4. Mitchell, R.; Williamson, A.M. Evaluation of an 8 hour versus a 12 hour shift roster on employees at a power station. *Appl. Ergon.* **2000**, *31*, 83–93. [CrossRef]

5. Takahashi, M.; Tanigawa, T.; Tachibana, N.; Mutou, K.; Kage, Y.; Smith, L.; Iso, H. Modifying effects of perceived adaptation to shift work on health, wellbeing, and alertness on the job among nuclear power plant operators. *Ind. Health* **2005**, *43*, 171–178. [CrossRef] [PubMed]

6. Murtezani, A.; Hundozi, H.; Orovcanec, N.; Berisha, M.; Meka, V. Low back pain predict sickness absence among power plant workers. *Indian J. Occup. Environ. Med.* **2010**, *14*, 49–53. [CrossRef]

7. Hole, J.; Pande, M. Worker productivity, occupational health, safety and environmental issues in thermal power plant. In Proceedings of the 2009 IEEE International Conference on Industrial Engineering and Engineering Management, Hong Kong, China, 8–11 December 2009; Volume 8, pp. 1082–1086. [CrossRef]

8. Sánchez, A.S.; Fernández, P.R.; Lasheras, F.S.; Juez, F.D.C.; Nieto, P.G. Prediction of work-related accidents according to working conditions using support vector machines. *Appl. Math. Comput.* **2011**, *218*, 3539–3552. [CrossRef]

9. Krzemień, A. Dynamic fire risk prevention strategy in underground coal gasification processes by means of artificial neural networks. *Arch. Min. Sci.* **2019**, *64*, 3–19. [CrossRef]

10. Krzemień, A. Fire risk prevention in underground coal gasification (UCG) within active mines: Temperature forecast by means of MARS models. *Energy* **2019**, *170*, 777–790. [CrossRef]

11. Asensio-Cuesta, S.; Diego-Mas, J.A.; Alcaide-Marzal, J. Applying generalised feedforward neural networks to classifying industrial jobs in terms of risk of low back disorders. *Int. J. Ind. Ergon.* **2010**, *40*, 629–635. [CrossRef]

12. Sánchez, A.S.; Iglesias-Rodríguez, F.; Fernández, P.R.; Juez, F.D.C. Applying the K-nearest neighbor technique to the classification of workers according to their risk of suffering musculoskeletal disorders. *Int. J. Ind. Ergon.* **2016**, *52*, 92–99. [CrossRef]

13. Serrano, N.M.B.; Nieto, P.J.G.; Sánchez, A.S.; Lasheras, F.S.; Fernández, P.R. A Hybrid Algorithm for the Assessment of the Influence of Risk Factors in the Development of Upper Limb Musculoskeletal Disorders. In *Lecture Notes in Computer Science*; Springer Science and Business Media LLC: Philadelphia, PA, USA, 2018; pp. 634–646.

14. Serrano, N.B.; Sánchez, A.S.; Lasheras, F.S.; Iglesias-Rodríguez, F.; Valverde, G.F. Identification of gender differences in the factors influencing shoulders, neck and upper limb MSD by means of multivariate adaptive regression splines (MARS). *Appl. Ergon.* **2019**, *82*, 102981. [CrossRef]

15. Ríos, E.M.A.; Sáchez-Lasheras, F.; Sánchez, A.S.; Iglesias-Rodríguez, F.J.; Crespo, M.D.M.S. Prediction of Computer Vision Syndrome in Health Personnel by Means of Genetic Algorithms and Binary Regression Trees. *Sensors* **2019**, *19*, 2800. [CrossRef]

16. Ríos, E.M.A.; Sánchez, A.S.; Sáchez-Lasheras, F.; Crespo, M.D.M.S. Genetic algorithm based on support vector machines for computer vision syndrome classification in health personnel. *Neural Comput. Appl.* **2018**, *32*, 1239–1248. [CrossRef]

17. Eurofound (European Foundation for the Improvement of Living and Working Conditions). *European Working Conditions Survey Integrated Data File, 1991–2015 [Data Collection]*, 7th ed.; UK Data Service: Colchester, UK, 2018.

18. UKDS (UK Data Service) Website. Available online: https://www.ukdataservice.ac.uk/ (accessed on 30 March 2020).

19. Eurofound. *Sixth European Working Conditions Survey—Overview Report*; Publications Office of the European Union: Luxembourg, 2016.
20. Eurofound Website. EWCS 2015—Source Questionnaire. Available online: https://www.eurofound.europa.eu/sites/default/files/page/field_ef_documents/6th_ewcs_2015_final_source_master_questionnaire.pdf (accessed on 24 April 2020).
21. Afifi, A.; Clark, V.; May, S. *Computer-Aided Multivariate Analysis*; Chapman & Hall/CRC: Boca Raton, FL, USA, 2004.
22. Bhattacharyya, H.T.; Kleinbaum, D.G.; Kupper, L.L. Applied Regression Analysis and Other Multivariable Methods. *J. Am. Stat. Assoc.* **1979**, *74*, 732. [CrossRef]
23. Ordóñez, C.; Sáchez-Lasheras, F.; Roca-Pardiñas, J.; Juez, F.J.D.C. A hybrid ARIMA–SVM model for the study of the remaining useful life of aircraft engines. *J. Comput. Appl. Math.* **2019**, *346*, 184–191. [CrossRef]
24. Juez, F.J.D.C.; Sáchez-Lasheras, F.; Roqueni, N.; Osborn, J. An ANN-Based Smart Tomographic Reconstructor in a Dynamic Environment. *Sensors* **2012**, *12*, 8895–8911. [CrossRef] [PubMed]
25. Holland, J.H. *Adaptation in Natural and Artificial Systems*; MIT Press – Journals: Cambridge, MA, USA, 1992.
26. Nieto, P.G.; Lasheras, F.S.; Juez, F.D.C.; Alonso-Fernández, J.R. Study of cyanotoxins presence from experimental cyanobacteria concentrations using a new data mining methodology based on multivariate adaptive regression splines in Trasona reservoir (Northern Spain). *J. Hazard. Mater.* **2011**, *195*, 414–421. [CrossRef]
27. De Andrés, J.; Sánchez-Lasheras, F.; Lorca, P.; De Cos Juez, F.J. A hybrid device of Self Organizing Maps (SOM) and Multivariate Adaptive Regression Splines (MARS) for the forecasting of firms' bankruptcy. *Account. Manag. Inf. Syst. Contab. Inform. Gestiune* **2011**, *10*, 351–374.
28. Alonso-Fernández, J.R.; Muñiz, C.D.; Nieto, P.G.; Juez, F.D.C.; Lasheras, F.S.; Roqueni, N. Forecasting the cyanotoxins presence in fresh waters: A new model based on genetic algorithms combined with the MARS technique. *Ecol. Eng.* **2013**, *53*, 68–78. [CrossRef]
29. Sprent, P.; Draper, N.R.; Smith, H. Applied Regression Analysis. *Biomaterials* **1981**, *37*, 863. [CrossRef]
30. Jakus, D.; Čađenović, R.; Vasilj, J.; Sarajčev, P. Optimal Reconfiguration of Distribution Networks Using Hybrid Heuristic-Genetic Algorithm. *Energies* **2020**, *13*, 1544. [CrossRef]
31. Krzywanski, J. A General Approach in Optimization of Heat Exchangers by Bio-Inspired Artificial Intelligence Methods. *Energies* **2019**, *12*, 4441. [CrossRef]
32. Hallman, D.M.; Holtermann, A.; Björklund, M.; Gupta, N.; Rasmussen, C.D.N. Sick leave due to musculoskeletal pain: Determinants of distinct trajectories over 1 year. *Int. Arch. Occup. Environ. Health* **2019**, *92*, 1099–1108. [CrossRef] [PubMed]
33. Benavides, F.G.; Benach, J.; Moncada, S.; Vahtera, J.; Kivimaki, M. Working conditions and sickness absence: A complex relation. *J. Epidemiol. Community Health* **2001**, *55*, 368. [CrossRef] [PubMed]
34. Hellig, T.; Rick, V.; Stranzenbach, R.; Przybysz, P.; Mertens, A.; Brandl, C. Investigation of the Effectiveness of European Assembly Worksheet in Assessing Organizational Measures for MSD Risk Assessment. In *Advances in Intelligent Systems and Computing*; Springer Science and Business Media LLC: Philadelphia, PA, USA, 2017; Volume 602, pp. 229–235.
35. Motamedzade, M.; Faghih, M.A.; Golmohammadi, R.; Faradmal, J.; Mohammadi, H. Effects of Physical and Personal Risk Factors on Sick Leave Due to Musculoskeletal Disorders. *Int. J. Occup. Saf. Ergon.* **2013**, *19*, 513–521. [CrossRef] [PubMed]
36. Kemmlert, K. Economic impact of ergonomic intervention—Four case studies. *J. Occup. Rehab.* **1996**, *6*, 17–32. [CrossRef]
37. Parenmark, G.; Malmkvist, A.-K.; Örtengren, R. Ergonomic moves in an engineering industry: Effects on sick leave frequency, labor turnover and productivity. *Int. J. Ind. Ergon.* **1993**, *11*, 291–300. [CrossRef]
38. Farhadi, R.; Omidi, L.; Balabandi, S.; Barzegar, S.; Abbasi, A.L.; Poornajaf, A.H.; Karchani, M. Investigation of musculoskeletal disorders and its relevant factors using quick exposure check (QEC) method among Seymareh hydropower plant workers. *J. Res. Health* **2014**, *4*, 715–720.
39. Mohanty, P.; Mohanty, S. Impact of Workplace Bullying on Performance, Psychological Distress and Absenteeism: An Original Review of Healthcare Sector. *J. Econ. Perspect.* **2017**, *11*, 1277–1286.
40. Petrén, V.; Petzäll, K.; Preber, H.; Bergstrom, J. The relationship between working conditions and sick leave in Swedish dental hygienists. *Int. J. Dent. Hyg.* **2007**, *5*, 27–35. [CrossRef]

41. Christmansson, M.; Fridén, J.; Sollerman, C. Task design, psycho-social work climate and upper extremity pain disorders—Effects of an organisational redesign on manual repetitive assembly jobs. *Appl. Ergon.* **1999**, *30*, 463–472. [CrossRef]

42. Lee, L.-K.; Yang, S.-M.; Park, J.; Kim, J. The Effort of Health Management for Workers in Y Combined Cycle Power Plant in Korea. *Toxicol. Environ. Health Sci.* **2018**, *10*, 42–48. [CrossRef]

43. Piha, K.; Laaksonen, M.; Martikainen, P.; Rahkonen, O.; Lahelma, E. Interrelationships between education, occupational class, income and sickness absence. *Eur. J. Public Health* **2009**, *20*, 276–280. [CrossRef] [PubMed]

44. Hansen, A.C.; Selte, H.K. Air Pollution and Sick-leaves. *Environ. Resour. Econ.* **2000**, *16*, 31–50. [CrossRef]

45. Lee, J.; Lee, W.; Choi, W.-J.; Kang, S.-K.; Ham, S. Association between Exposure to Extreme Temperature and Injury at the Workplace. *Int. J. Environ. Res. Public Health* **2019**, *16*, 4955. [CrossRef]

46. Huijs, J.J.J.M.; Koppes, L.L.J.; Taris, T.W.; Blonk, R.W.B. Differences in Predictors of Return to Work Among Long-Term Sick-Listed Employees with Different Self-Reported Reasons for Sick Leave. *J. Occup. Rehab.* **2012**, *22*, 301–311. [CrossRef]

47. Eriksen, W.; Bruusgaard, D. Physical Leisure-Time Activities and Long-Term Sick Leave: A 15-Month Prospective Study of Nurses Aides. *J. Occup. Environ. Med.* **2002**, *44*, 530–538. [CrossRef]

48. Hildebrandt, V.H.; Bongers, P.M.; Dul, J.; Van Dijk, F.J.H.; Kemper, H.C.G. The relationship between leisure time, physical activities and musculoskeletal symptoms and disability in worker populations. *Int. Arch. Occup. Environ. Health* **2000**, *73*, 507–518. [CrossRef]

49. Montano, D. A psychosocial theory of sick leave put to the test in the European Working Conditions Survey 2010–2015. *Int. Arch. Occup. Environ. Health* **2019**, *93*, 229–242. [CrossRef]

Article

Energy Multiphase Model for Biocoal Conversion Systems by Means of a Nodal Network

Beatriz M. Paredes-Sánchez [1], José P. Paredes-Sánchez [1,*] and Paulino J. García-Nieto [2]

[1]   Department of Energy, College of Mining, Energy and Materials Engineering, University of Oviedo, 33004 Oviedo, Spain; uo19070@uniovi.es
[2]   Department of Mathematics, Faculty of Sciences, University of Oviedo, 33007 Oviedo, Spain; pjgarcia@uniovi.es
*   Correspondence: paredespablo@uniovi.es

Received: 28 April 2020; Accepted: 25 May 2020; Published: 29 May 2020

Abstract: The coal-producing territories in the world are facing the production of renewable energy in their thermal systems. The production of biocoal has emerged as one of the most promising thermo-energetic conversion technologies, intended as an alternative fuel to coal. The aim of this research is to assess how the model of biomass to biocoal conversion in mining areas is applied for thermal systems engineering. The Central Asturian Coal Basin (CACB; Spain) is the study area. The methodology used allows for the analysis of the resource as well as the thermo-energetic conversion and the management of the bioenergy throughout the different phases in a process of analytical hierarchy. This is carried out using a multiphase mathematical algorithm based on the availability of resources, the thermo-energetic conversion, and the energy management in the area of study. Based on the working conditions, this research highlights the potential of forest biomass as a raw material for biocoal production as well as for electrical and thermal purposes. The selected node operates through the bioenergy-match mode, which has yielded outputs of 23 MW_e and 172 MW_{th}, respectively.

Keywords: biomass; bioenergy; energy production system

1. Introduction

The constant technological progress and the increasing industrialization of society have boosted the demand for energy. Fossil fuel deposits, such as those in coal basins, are limited and have been extensively exploited, so modelling the biomass potential of renewable sources in these areas is a challenge for the future in the European Union (EU). Spain boasts a wide variety of renewable resources such as biomass and wind and solar energy, all available as renewable energy sources [1]. The technological capacity of Spanish industry has made it a benchmark in the use of renewable resources [2]. The potential of Spain in renewable energies is well above both the domestic energy demand and the existing fossil fuel resources [3]. Despite this situation, Spain is highly dependent on foreign energy from fossil fuels, coal being the main source of indigenous energy in Spain [4].

Renewable energy sources are found in nature, have the capacity to be totally or partially regenerated, and can be used for energy purposes. Biomass is part of a continual cycle of mass and energy consumption and production in the environment. It can be used to produce energy, either directly by combustion or indirectly by way of biofuels. EU Directive 2003/30/EC [5] defines biomass as a biodegradable fraction of products, waste and residues from agriculture, including vegetal and animal substances, forestry and related industries, as well as the biodegradable fraction of industrial and municipal waste. Such a definition has a comprehensive character since it includes a variety of energy sources sharing certain characteristics but differing in their origin and the technology used to obtain and use them, one of the main sources being forest biomass.

The EU Forestry Action Plan [6] calls for an assessment of forest biomass availability for energy production in both national and regional energy systems. The most promising scenarios for future use are

1. Production of electricity either by co-combustion or direct combustion;
2. Thermal applications for domestic or industrial consumers;
3. Production of solid biofuels to be used in the cement or steel industry.

This study is to be carried out in the Region of Asturias in northern Spain, a region rich in fossil fuels, i.e., coal, where the production of alternative biofuels has been of interest in the context of the EU [7,8]. Its producing industry uses coal as the main source of indigenous primary energy, a part of which is produced in its mining basins [9]. Mining has gradually declined in recent years following the reduction of its activity [10]. Biomass from forests and forest management in the mining areas is conditioned upon its frequency of occurrence along with its topography. It is, therefore, necessary to carefully plan management procedures based on the distribution of potentials throughout the whole territory. This strategic framework calls for an analysis of the most appropriate techniques for extraction, storage, transport and use of biomass in forests. However, there is limited use of forest biomass to provide an alternative to fossil fuels owing to its low energy density [11].

Biocoal is achieved through a roasting process (i.e., torrefaction process), a high potential process to be applied in the production of solid biofuels from lignocellulosic biomass [12]. In this way, the quality of this fuel increases by virtue of its hydrophobicity and higher calorific value compared to the initial biomass [13]. Biofuels considerably improve the potential of biomass for industry and thermal systems [14]. Furthermore, it is of interest for both industrial energy transition and industry 4.0 [15]. However, modelling, conversion and energy management of resource supply are some of the main challenges in the current energy context [16]. In this regard, Visa et al. [17] have defined the importance of characterizing energy production systems according to the nature of the energy resource used. Paredes-Sánchez and Ochoa-Lopez [18] established the importance of biomass modelling as an alternative to coal in the energy production systems. Therefore, a proper implementation in a traditionally coal-based industry, as is the case of the Principality of Asturias, requires the development of mathematical models for the energy conversion of its resources.

Extensive research has been carried out using both resource characterization and energy conversion models from different perspectives [19–22]. These current works discuss numerous challenging issues, including the increasing number of assumptions that ensure consistency between the models used and the large amount of data required for energy conversion. The combination of modelling methods allows for a significant number of variables to be taken into account to solve the problems of energy conversion [22]. The multiphase model is an important step in developing the knowledge needed to improve energy fuels in thermal systems. In this context, research and development activities on biomass torrefaction have been particularly active in exploring its potential as a fuel [23,24]. However, the modelling and management of the torrefaction process can be found in the literature as a current challenge [25,26]. Kumar et al. [27] have studied the processes of bioenergy transformation, pointing out the necessity to search for comprehensive analytical models to overcome the existing limitations and provide the necessary technology to enable industrial use of high value-added biofuels. Huntington et al. [28] stress the importance of developing advanced mathematical models focused on the potential supply of biomass resources as a source of bioenergy. Paredes-Sánchez et al. [29] point out the limitations in modelling the potential use of biofuels for energy production systems considering the scope of the energy conversion system. In this regard, Bach et al. [30] define the demand for a comprehensive model of biomass torrefaction, which can provide interdisciplinary information to industrialize and commercialize the process. In this framework, a comprehensive analysis of the energy use of biocoal, starting with the supply of biomass as raw material and ending with its final energy conversion through fossil-fuel-based technologies for thermal systems, is a difficult undertaking in the field of bioenergy. Therefore, the present paper aims to develop a mathematical model that would allow a comprehensive analysis of the potential use of available biomass in the mining areas to produce biocoal for thermal systems engineering. The main breakthrough consists of applying a multiphase model to the data to characterize thermo-energetic conversion and energy management of biomass as a raw material for the production of biocoal. These studies are carried out in the Carboniferous Basin of the Principality of Asturias (CACB).

This work is organized as follows: Section 1 introduces the context of the research and the aim of the work. Section 2 shows the methodology used for the study area and the modelling process to produce biocoal. Section 3 shows both its findings and the details of its implementation as results. Section 4 provides a more detailed analysis of and discussion on biocoal as an alternative fuel to coal in mining areas for energy production systems. Section 5 provides the main conclusions of the study in the mining area by means of energy production systems.

2. Materials and Methods

2.1. Study Area and Mine Nodes

The National Renewable Energy Action Plan (NREAP) in Spain is aimed at complying with European Directives 2009/28/EC and 2009/29 EC on the contribution of renewable energy and the reduction of greenhouse gas (GHG) emissions by 2020 [3]. Located in the north of Spain, the Principality of Asturias is an Autonomous Community where 45% of the land is forested. The energy structure of the Principality of Asturias is conditioned by the contribution of fossil fuels to the national energy system as a whole. The Central Asturian Coal Basin (CACB) spreads over the southern councils of the central area of the region (Figure 1).

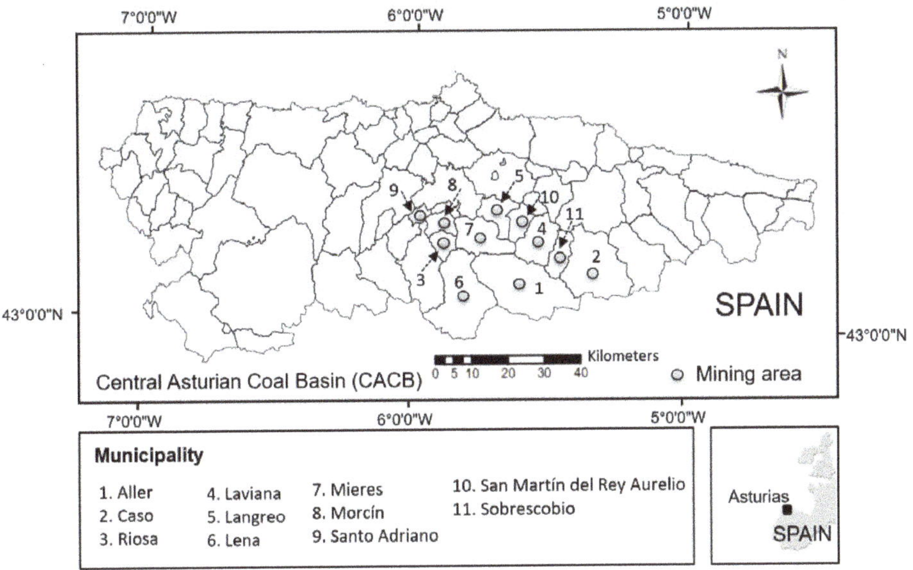

Figure 1. Location of the study area—Principality of Asturias (Spain).

The CACB features the geological resources that represent one of the main coal mining deposits in Spain, located above the geological unit of study [31]. The use of coal as an integral source of indigenous energy is the basis of the Asturian economy. Therefore, the CACB constitutes a geographical, economic, and geological unit that is the study area. In this sense, mine shafts are defined as candidate nodes of analysis to implement the use of biocoal in energy production systems. It should be noted that the layout and infrastructure of the coal mine itself, i.e., the electricity grid or the transport network, would favor the use of biocoal as an alternative fuel. Table 1 shows the mine shafts considered as modelling nodes.

Table 1. Nodes defined in the Central Asturian Coal Basin (CACB) study area.

Code	Name of the Node	Municipality
PM	Pozo María Luisa-Samuño	Langreo
PSo	Pozo Sotón	San Martín del Rey Aurelio
PSa	Pozo Santiago	Aller
PMo	Pozo Montsacro	Riosa
PS	Pozo San Nicolás	Mieres
PC	Pozo Candín	Langreo
PCa	Pozo Carrio	Laviana

The analysis of the biomass in the forests surrounding the mine shafts allows unused resources to be valorized for energy production. The use of forest biomass for energy purposes will not only increase the economic development, but also energy self-sufficiency [32,33]. The use of forest residues will also generate environmental benefits [34].

2.2. Multiphase Mathematical Model

The multiphase mathematical model developed for this work comprises three phases in the area of study: resources, thermo-energetic conversion and energy management, each one of which is described in detail in Figure 2.

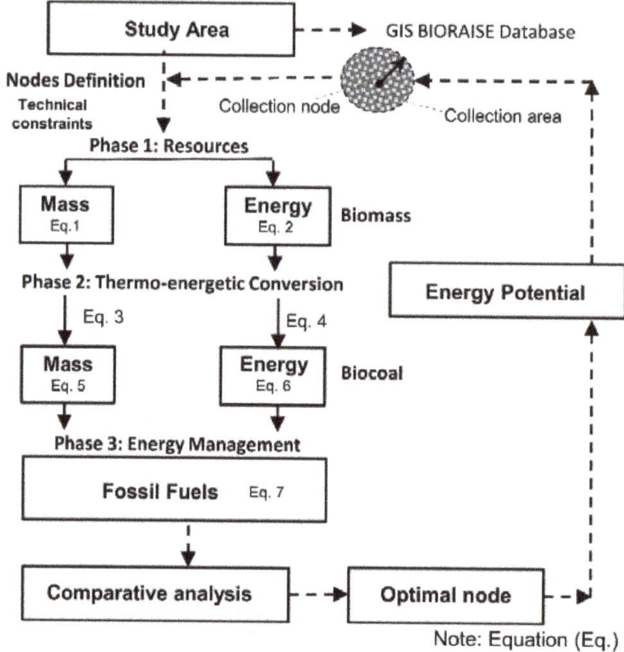

Figure 2. Flowchart of the multiphase mathematical algorithm through mass and energy balance.

2.2.1. Phase 1: Resources

For the assessment of potential resources, the Geographic Information System (GIS) database of the BIORAISE GIS tool from the Research Centre for Energy, Environment and Technology (CIEMAT) was used in order to collect data on forest biomass in the CACB study area [35]. In Phase 1 of the mathematical model, it is estimated that the existing resources for each candidate node come from residues of cleaning activities and forest management operations in the study area and are gathered in each of the considered nodes (Figure 2). Mass and energy are expressed in terms of dry ton (dry t). In this study, this biomass is considered as the raw material to be used to generate biocoal through the torrefaction technology in Phase 2 of the model with the roasting system.

The potential mass (M) is the residual forestry biomass, such as branches and leaves, found within the study area and therefore, in the surroundings of each modelling node. The available mass (m) consists of branches and tops (including leaves). It is obtained from cleaning, thinning and felling operations from BIORASE GIS database (Figure 2), which takes into account the techno-economic constraints of the potential biomass to define the useful resources [35]. Such restrictions derive from harvesting procedures, which depend on the terrain conditions to access the raw material [7] that will define the useful resources, including available mass of conifers, wood and mixtures (dry t/year). Techno-economic constraints affect its use from the point of view of mass and energy balance. The biomass is evaluated in the surroundings of the selected node in the study area using Equation (1).

$$m = \sum_{i=1}^{n} m_i \tag{1}$$

where

m is total available mass (dry t/year); and
m_i is total available mass of conifers, hardwood and mixtures (dry t/year).

The energy of the available residues (E) is the result of Equation (2), where the Lower Heating Value (LHV) is used.

$$E = \sum_{i=1}^{n} (m_i \cdot LHV_i) \tag{2}$$

where

E is energy from available mass (GJ/year);
m_i is total available mass of conifers, hardwood and mixtures (dry t/year); and
LHV_i is Lower Heating Value of conifers, hardwood and mixtures (GJ/dry t).

Phase 2 of the calculation algorithm is defined from the available resource and energy (Figure 2).

2.2.2. Phase 2: Thermo-Energetic Conversion

In Phase 2, the thermo-energetic calculation algorithm used to calculate the mass and energy potential of biocoal to be produced is based on Equations (1) and (2), respectively. Torrefaction is a thermal pre-treatment carried out by a reactor system at atmospheric pressure, with no oxygen. The roasting process takes place at temperatures between 200 °C and 300 °C to achieve more uniform solid biofuel, whose two main thermo-energetic characteristics are energy yield and mass yield. The energy and mass yields [36] are defined from the reactive part of the biomass, which is turned into biocoal through torrefaction as energy and mass percentages or fractions [36,37]; therefore, both ash and free water content are excluded from the definition. The mass yield (α_y) is the correlation between the mass of the biocoal produced (m_b) and the mass of the raw material, i.e., total available mass (m), obtained from the resources in the roasting system per candidate node (Equation (3)).

$$\alpha_y = \left(\frac{m_b}{m} \right) \tag{3}$$

Energy yield (β_y) is defined in the model according to Equation (4)

$$\beta_y = \alpha_y \cdot \left(\frac{LHV_b}{LHV}\right) \tag{4}$$

where

"β_y" is the yield referred to the LHV parameter already considered in Phase 1;
"α_y" is mass yield; and
LHV_b and LHV are lower heating value of biocoal mass and raw material, respectively [36].

The available energy per unit of mass corresponds to LHV because it represents the energy that can be efficiently recovered after combustion. To this objetive, the torrefaction process is assumed to be carried out on the available mass as a whole within the areas surrounding each node, thus obtaining Equation (5):

$$m_b = \left(\sum_{i=1}^{n} m_i\right)\cdot\alpha_y \tag{5}$$

where

m_b is biocoal mass from available mass per node (dry t/year);
m_i is mass of available biomass (dry t/year); and
α_y is mass yield.

For the calculation of energy potential to be obtained as primary energy from biocoal, Equation (6) is applied as an approach.

$$E_b = \sum_{i=1}^{n} (m_i\cdot LHV_i)\cdot\beta_y \tag{6}$$

where

E_b is bioenergy as biocoal per node (GJ/year);
m_i is available mass (dry t/year);
LHV_i is lower heating value of conifers, hardwood and mixtures (GJ/dry t); and
β_y is energy yield.

Equations (3) and (4) show the technical feasibility of placing a roasting system at a candidate node in the study area by bioenergy-match mode. Such operating conditions correspond to short residence periods, less than 30 min, with temperatures above 260 °C. Favorable conditions are considered to be those with energy and mass efficiency above 95% and 90% respectively. If the initial biomass to be torrefacted is dry, with moisture content below 10–15%, lower energy efficiency can be expected [37].

2.2.3. Phase 3: Energy Management

Biomass represents the main manageable renewable energy and should therefore play a role in energy production to replace or complement fossil fuels (Phase 3). Introducing biocoal in the industrial sector, and especially in thermal demand, calls for a change in the energy supply management model [38]. Energy production systems consist of sets of technologies that transform raw energy from one fuel into final use energy [17], allowing for the production of heat and/or electricity and of new biofuels. Based on the analysis of energy management, the energy to be produced will be assessed on the basis of the biocoal production in the roasting system as compared to its energy-based equivalence to the mass (m_f) of other conventional fuels, Equation (7).

$$m_f = E_b / CV_i \tag{7}$$

This is done by taking the calorific value (CV_i) of the fuels in Table 2 [39] as an approximation.

Table 2. Energy-based equivalence of certain fuels.

Type of Fuel	Calorific Value (CV$_i$) (GJ/t)
Brown coal	18.8
Distilled oil	41.2
Natural gas	45.6

The model of analysis using the multiphase mathematical algorithm considers a maximum distance of 50 km for the resources around each candidate node, which is a limit distance determined for their transformation into conventional biofuels that can be used in that territory [40]. Consequently, the calculation area is restricted to a maximum of about 50 km away from each candidate node for the entire analysis (Phases 1, 2, and 3). Once the final combination of the results per candidate node has been determined, the optimal node, i.e., the location of the roasting system in one mine shaft, will be defined for the harvesting of the available resources in the study area.

3. Results

In Phase 1, the analysis model established by the multiphase mathematical algorithm (Figure 2) shows the quantities of both potential biomass and available biomass in the study area for each of the nodes considered, as shown in Figure 3.

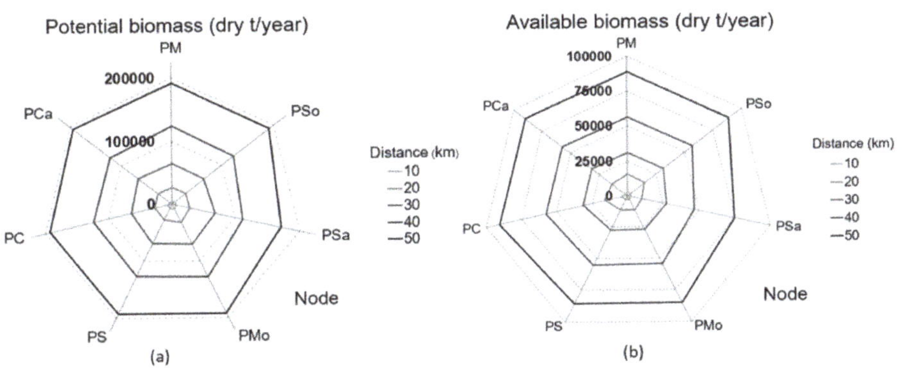

Figure 3. Mass per candidate node in the study area: (**a**) potential biomass (M) and (**b**) available biomass (m).

Figure 3 shows some linearity in the distribution of the potential biomass around each modelling node considered in the CACB. It indicates a regular distribution of the biomass with the increase of the distance from 10 km to 50 km. Increasing the distance yields more available biomass around each candidate node (Equation (1)). The available energy based on the available biomass per candidate node is shown in Figure 4.

For each node, the amounts of the available biomass that can be collected are above 75 dry kt/year, around 1300 TJ/year. The available biomass will be used as raw material to be converted into biocoal (Figure 4).

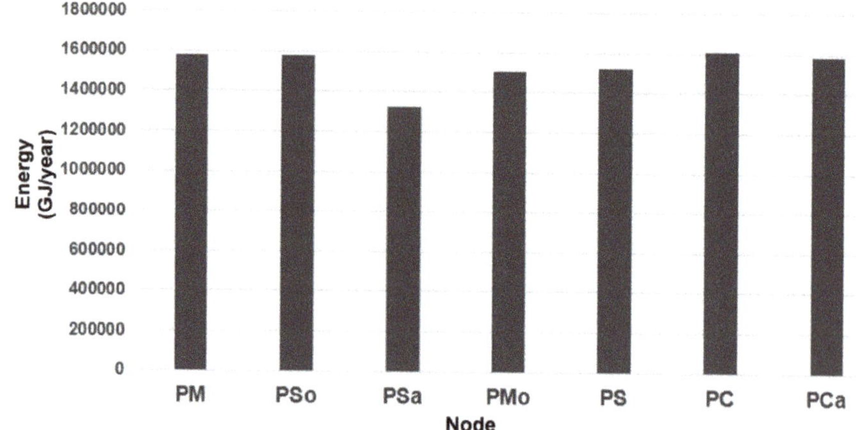

Figure 4. The energy (E) of the available biomass per candidate node as calculated by Equation (2).

Phase 2 takes into account the thermo-energetic conversion of the available biomass into biocoal as per Equations (5) and (6). Figure 5 shows the results of the biocoal potential and its equivalent energy per modelling node with the increase of the distance from 10 km to 50 km. The thermo-energetic conversion does not affect the distribution because it has a similar shift of the mass and energy per candidate node in the study area.

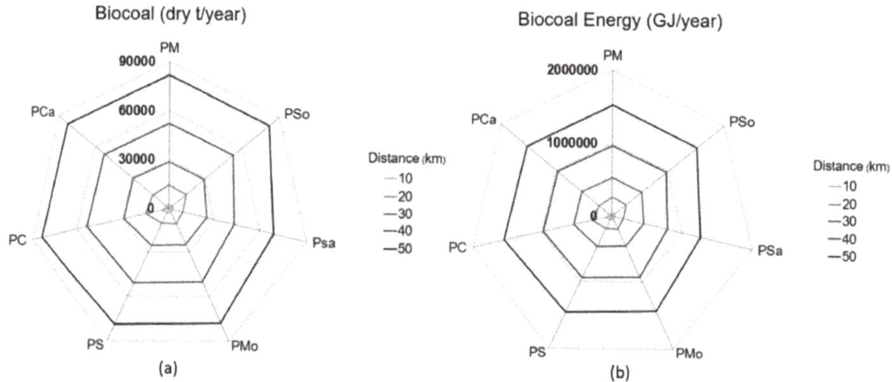

Note: distance is defined in radial axis.

Figure 5. Mass and energy per candidate node in the study area: (**a**) biocoal (m_b), (**b**) bioenergy (E_b).

Furthermore, the LHV distribution of the biocoal obtained from the raw material collected in the study area has been evaluated per candidate node (Figure 6).

As can be seen, there is a homogeneous distribution of the energy quality of the theoretical biocoal to be obtained, reaching an overall average value of about 18.5 GJ/dry t.

Finally, Phase 3 of the algorithm is implemented in each node, based on the data in Table 2. Equation (7) shows this potential as an alternative fuel to coal for energy conversion. This assessment characterizes biocoal as compared to different types of fossil fuels for energy production systems (Figure 7).

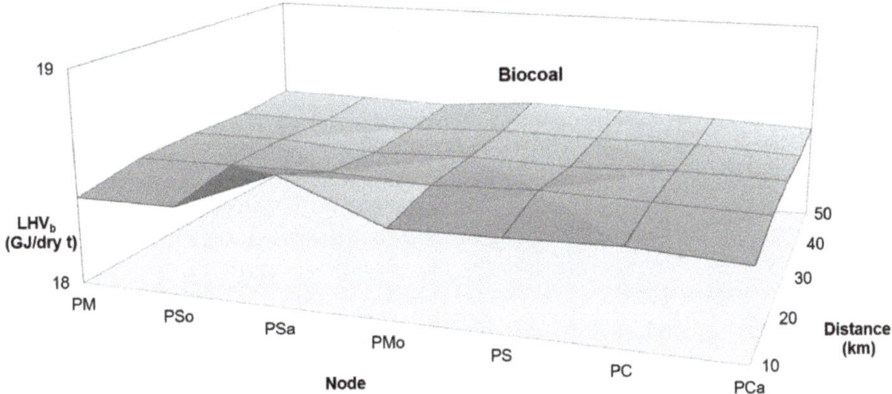

Figure 6. Lower heating value (LHV$_b$) distribution of the obtainable biocoal in the study area.

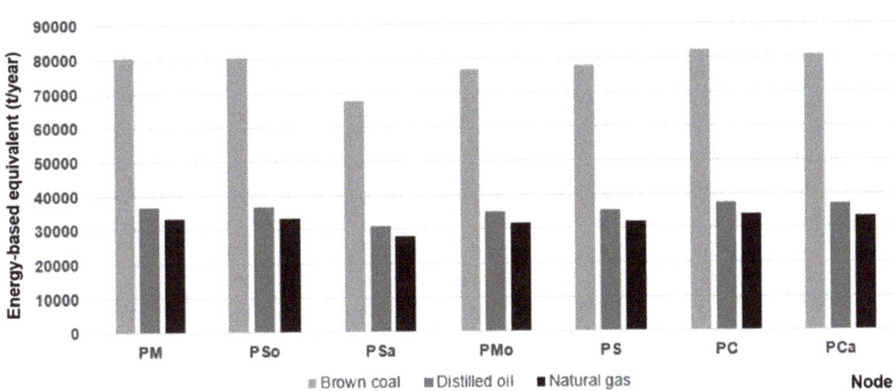

Figure 7. Energy-based equivalent fuels from biocoal, as a substitute in energy production systems.

Taking all candidate nodes, the potentially producible biocoal in terms of energy is equivalent of above 28 kt/year of natural gas, 31 kt/year of distilled oil, and 68 kt/year of brown coal.

4. Discussion

Implementing a technological alternative requires splitting up the analysis by considering the different aspects involved [41]. The paper explores biocoal use as an alternative to coal production in mining basins by using mine shafts as energy conversion nodes that draw on the surrounding forest biomass as raw material. The findings of the study reveal that the proposed model, based on a multiphase algorithm, is highly useful in the study of energy conversion and management for thermal systems. A cumulative analysis of these facts, augmented by other findings of this work, is summarized below.

The potential renewable in the world is diverse and widely distributed in nature [1]. However, most of it may present supply fluctuations, giving rise to risks and uncertainties in serving regional energy markets. In this sense, the use of biocoal in mining environments, as proposed, guarantees the use of renewable resources and their conversion, since it is based on substituting the use of fuel, i.e., coal, with one of similar characteristics, i.e., biocoal. Moreover, there is a whole industrial structure with mature technologies for the use of this coal. Despite this, one of the remaining challenges for the development of biofuels in the industry is whether a sufficient amount of raw material can be generated to meet the growing demand and achieve a shift from fossil fuels to biofuels [42].

In this regard, the applied algorithm shows that any node considered as a thermo-energetic conversion node for biocoal production has at least 75 dry kt/year of raw material for biocoal production in the study area. Here, given the nature of the area under consideration and the nodes, only one node may be selected, given the territorial extension of the CACB.

Biocoal could replace coal in the production of heat and power through energy production systems. It would also have a large impact on industry [43]. Providing the infrastructure, technology and research to enable the use of renewable energy sources available "on site" is a challenge for industry, in particular for readily available sources such as biomass [38]. The continued development of industrial technology is expected that will ensure more competitive, reliable and sustainable energy production systems [44]. The use of biomass as an alternative to coal in certain contexts has been made feasible by using biocoal technology, but advances in technology alone do not promote its widespread use; therefore, extensive studies are needed. In this sense, from the mass and energy balance of the analysis model, the optimal node appears to be PC, this is selected as the optimal node for the CACB. With the use of a roasting system, this node would produce about 84 kt/year of biocoal, equivalent to 1544 TJ/year of biocoal, the largest amount of all the nodes selected in the proposed analysis.

Renewable energy generation systems are illustrative of the fact that some renewable energy infrastructures are at a significant distance from energy conversion systems [41]. Overcoming this barrier to competitiveness of renewable energies requires a well-planned and carefully managed energy supply infrastructure during and after the infrastructure investment. An environment based on fuel-coal energy production systems in the mining basins benefits both their use and direct conversion for electrical, thermal or industrial purposes, given the development of a coal energy conversion industry in the coal mining environment. However, all this requires some analysis of future techno-economic feasibility as well as in-depth studies based on the development of these technologies.

Biomass holds the potential to be considered the best alternative to meet global energy demand sustainably and reduce the impact of polluting energy resources [45]. In this context, the development of biorefineries for the production of advanced biofuels is encouraged. Biorefinery represents a sustainable means of generating multiple bioenergy products from various biomass raw materials via the incorporation of relevant conversion technologies. Biorefinery is crucial in the transition of various traditional industries into a circular bioeconomy in the context of energy transition [46]. That opens up the door to the development of infrastructures such as the biorefinery in the aforementioned node. The use of biomass in energy production systems such as co-firing coal with biomass to generate energy is gradually increasing even though their performance differs significantly due to the wide variations in its physical and chemical properties [47]. However, biocoal overcomes most of its use limitations for heat or electricity production.

In this context, in the shaft named "Pozo Candín" (PC), considering the efficiency of a heat production system with a thermal efficiency of 80% and about 2000 h of operation per year [48], it is possible to install a total power of 172 MW_{th} for the energy conversion of biocoal. Additionally, if it is considered the alternative objective of electricity production, either by co-combustion or by itself, looking at an electricity efficiency in the complete conversion of 40% for about 7500 h, it would be possible to achieve an electricity potential equivalent to 23 MW_e [49,50]. Detailed studies of biomass conversion stages within the energy production system are working lines for the future in order to further develop more efficient and advanced energy conversion systems.

5. Conclusions

Biocoal makes it possible to overcome many of the barriers that condition its use and the development of biomass in mining basins around the world due to the traditional interrelationship between the resource and conversion technology for its viability. This is an opportunity to focus on using it in energy production systems that use coal as an energy source, as it overcomes many of the limitations traditionally associated with biomass. In this respect, a multiphase mathematical algorithm based on operational resource data, thermo-energetic conversion, and energy management has been developed.

The research in this paper highlights the potential of forest biomass in mining areas around the world, in this case in the CACB in Spain, and identifies possible options for the use of forest residues as raw material for solid biofuels as an alternative to coal in carbon basins, i.e., biocoal. Taking a minimum value considering all the nodes in the study area, the energy potential for biocoal is equivalent to over 28 kt/year of natural gas, 31 kt/year of distilled oil, and 68 kt/year of brown coal.

Overall, the quantities of biomass that can be collected in all the nodes studied are above 75 dry kt/year (about 1300 TJ/year). The optimal node that guarantees the largest amount of energy, from a point of view of the proposed mass and energy balance in the study area, is the "Pozo Candín" (PC), which would allow biocoal to be produced through a roasting system at a rate of 84 kt/year, equivalent to 1544 TJ/year in the considered operation conditions with the biomass of the CACB. The production potential of electrical and thermal energy by thermal systems under the defined conditions in this node amounts to 23 MW_e and 172 MW_{th}, respectively.

Implementing this energy potential on an industrial scale in the CACB requires techno-economic and thermal systems studies based on the specific characteristics and objectives of the facilities that will be using it. Future understanding of the range of benefits and challenges when introducing and up-scaling biocoal production under different scenarios will depend on detailed, comprehensive, and simultaneous assessments, technological options, and final techno-economic factors.

Author Contributions: Conceptualization and methodology, B.M.P.-S. and J.P.P.-S.; software, J.P.P.-S.; writing—original draft preparation, B.M.P.-S.; writing—review and editing, B.M.P.-S.; visualization, J.P.P.-S. and P.J.G.-N.; supervision, B.M.P.-S., J.P.P.-S., and P.J.G.-N. All authors have read and agreed to the published version of the manuscript.

Funding: This research received no external funding. It is an initiative of research through an experimental software implementation for technology transfer in thermal systems in the University of Oviedo.

Acknowledgments: The authors would like to acknowledge the support of the technical staff of the Research Centre for Energy, Environment and Technology (CIEMAT), throughout the various stages of this work.

Conflicts of Interest: The authors declare no conflict of interest.

References

1. Heras-Saizarbitoria, I.; Sáez, L.; Allur, E.; Morandeira, J. The emergence of renewable energy cooperatives in Spain: A review. *Renew. Sustain. Energy Rev.* **2018**, *94*, 1036–1043. [CrossRef]

2. Sánchez-Durán, R.; Luque, J.; Barbancho, J. Long-term demand forecasting in a scenario of energy transition. *Energies* **2019**, *12*, 3095. [CrossRef]

3. MITECO (Ministerio para la Transición Ecológica y el Reto Demográfico). Plan de Energías Renovables (PER) 2011–2020, Primera Parte. (In Spanish). Available online: https://www.miteco.gob.es/es/cambio-climatico/legislacion/documentacion/PER_2011-2020_VOL_I_tcm30-178649.pdf (accessed on 18 March 2020).

4. MITECO (Ministerio para la Transición Ecológica y el Reto Demográfico). La Energía en España 2017. (In Spanish). Available online: https://energia.gob.es/balances/Balances/LibrosEnergia/Libro-Energia-2017.pdf (accessed on 18 March 2020).

5. Directive 2003/30/EC of the European Parliament and of the Council of 8 May 2003 on the Promotion of the Use of Biofuels or Other Renewable Fuels for Transport. Available online: https://eur-lex.europa.eu/legal-content/EN/ALL/?uri=CELEX%3A32003L0030 (accessed on 19 March 2020).

6. European Union Forest Action Plan. Available online: https://eur-lex.europa.eu/legal-content/EN/TXT/?uri=LEGISSUM%3Al24277 (accessed on 18 March 2020).

7. Paredes-Sánchez, J.P.; Gutiérrez-Trashorras, A.J.; Xiberta-Bernat, J. Wood residue to energy from forests in the Central Metropolitan Area of Asturias (NW Spain). *Urban For. Urban Green.* **2015**, *14*, 195–199. [CrossRef]

8. Paredes-Sánchez, J.P.; Gutiérrez-Trashorras, A.J.; Xiberta-Bernat, J. Energy potential of residue from wood transformation industry in the central metropolitan area of the Principality of Asturias (northwest Spain). *Waste Manag. Res.* **2014**, *32*, 241–244. [CrossRef] [PubMed]

9.   SADEI (Sociedad Asturiana de Estudios Industriales). Datos Básicos de Asturias 2019 (In Spanish). Available online: http://www.sadei.es/datos/catalogo/m00/dabaas/2019/datos-basicos-asturias-2019.pdf (accessed on 19 March 2020).

10.   Moreno, B.; López, A.J. The effect of renewable energy on employment. The case of Asturias (Spain). *Renew. Sustain. Energy Rev.* **2008**, *12*, 732–751. [CrossRef]

11.   Tumuluru, J.S.; Wright, C.T.; Kenney, K.L.; Hess, J.R. *A Technical Review on Biomass Processing: Densification, Preprocessing, Modeling, and Optimization*; Idaho National Laboratory (INL): Idaho Falls, IA, USA, 2010.

12.   Dai, L.; Wang, Y.; Liu, Y.; Ruan, R.; He, C.; Yu, Z.; Jiang, L.; Zeng, Z.; Tian, X. Integrated process of lignocellulosic biomass torrefaction and pyrolysis for upgrading bio-oil production: A state-of-the-art review. *Renew. Sustain. Energy Rev.* **2019**, *107*, 20–36. [CrossRef]

13.   Batidzirai, B.; Mignot, A.P.R.; Schakel, W.B.; Junginger, H.M.; Faaij, A.P.C. Biomass torrefaction technology: Techno-economic status and future prospects. *Energy* **2013**, *62*, 196–214. [CrossRef]

14.   Niu, Y.; Lv, Y.; Lei, Y.; Liu, S.; Liang, Y.; Wang, D.; Hui, S. Biomass torrefaction: Properties, applications, challenges, and economy. *Renew. Sustain. Energy Rev.* **2019**, *115*, 109395. [CrossRef]

15.   Oztemel, E.; Gursev, S. Literature review of Industry 4.0 and related technologies. *J. Intell. Manuf.* **2020**, *31*, 127–182. [CrossRef]

16.   Nunes, L.J.R.; Causer, T.P.; Ciolkosz, D. Biomass for energy: A review on supply chain management models. *Renew. Sustain. Energy Rev.* **2020**, *120*, 109658. [CrossRef]

17.   Visa, I.; Duta, A.; Moldovan, M.; Burduhos, B.; Neagoe, M. *Solar Energy Conversion Systems in the Built Environment. Green Energy and Technology*; Springer: Cham, Switzerland, 2020; pp. 59–158.

18.   Paredes-Sánchez, J.P.; López-Ochoa, L.M. Bioenergy as an Alternative to Fossil Fuels in Thermal Systems. In *Advances in Sustainable Energy*; Vasel-Be-Hagh, A., Ting, D., Eds.; Springer: Cham, Switzerland, 2019; pp. 149–168.

19.   Murele, O.C.; Zulkafli, N.I.; Kopanos, G.; Hart, P.; Hanak, D.P. Integrating biomass into energy supply chain networks. *J. Clean. Prod.* **2020**, *248*, 119246. [CrossRef]

20.   Azevedo, S.G.; Sequeira, T.; Santos, M.; Mendes, L. Biomass-related sustainability: A review of the literature and interpretive structural modeling. *Energy* **2019**, *171*, 1107–1125. [CrossRef]

21.   García Nieto, P.J.; García-Gonzalo, E.; Sánchez Lasheras, F.; Paredes-Sánchez, J.P.; Riesgo Fernández, P. Forecast of the higher heating value in biomass torrefaction by means of machine learning techniques. *J. Comput. Appl. Math.* **2019**, *357*, 284–301. [CrossRef]

22.   Aalto, M.; Raghu, K.C.; Korpinen, O.J.; Karttunen, K.; Ranta, T. Modeling of biomass supply system by combining computational methods—A review article. *Appl. Energy* **2019**, *243*, 145–154. [CrossRef]

23.   Cahyanti, M.N.; Doddapaneni, T.R.K.C.; Kikas, T. Biomass torrefaction: An overview on process parameters, economic and environmental aspects and recent advancements. *Bioresour. Technol.* **2020**, *301*, 122737. [CrossRef] [PubMed]

24.   Hu, J.; Song, Y.; Liu, J.; Evrendilek, F.; Buyukada, M.; Yan, Y.; Li, L. Combustions of torrefaction-pretreated bamboo forest residues: Physicochemical properties, evolved gases, and kinetic mechanisms. *Bioresour. Technol.* **2020**, *304*, 122960. [CrossRef] [PubMed]

25.   Nunes, L.J.; Matias, J.C. Biomass Torrefaction as a Key Driver for the Sustainable Development and Decarbonization of Energy Production. *Sustainability* **2020**, *12*, 922. [CrossRef]

26.   Ribeiro, J.M.C.; Godina, R.; Matias, J.C.D.O.; Nunes, L.J.R. Future perspectives of biomass torrefaction: Review of the current state-of-the-art and research development. *Sustainability* **2018**, *10*, 2323. [CrossRef]

27.   Kumar, B.; Bhardwaj, N.; Agrawal, K.; Chaturvedi, V.; Verma, P. Current perspective on pretreatment technologies using lignocellulosic biomass: An emerging biorefinery concept. *Fuel Process. Technol.* **2020**, *199*, 106244. [CrossRef]

28.   Huntington, T.; Cui, X.; Mishra, U.; Scown, C.D. Machine learning to predict biomass sorghum yields under future climate scenarios. *Biofuel Bioprod. Biorefining* **2020**, *14*, 14. [CrossRef]

29.   Paredes-Sánchez, J.P.; Conde, M.; Gómez, M.A.; Alves, D. Modelling hybrid thermal systems for district heating: A pilot project in wood transformation industry. *J. Clean. Prod.* **2018**, *194*, 726–734. [CrossRef]

30.   Bach, Q.V.; Skreiberg, Ø.; Lee, C.J. Process modeling and optimization for torrefaction of forest residues. *Energy* **2017**, *138*, 348–354. [CrossRef]

31. Piedad-Sánchez, N.; Suárez-Ruiz, I.; Martínez, L.; Izart, A.; Elie, M.; Keravis, D. Organic petrology and geochemistry of the Carboniferous coal seams from the Central Asturian Coal Basin (NW Spain). *Int. J. Coal Geol.* **2004**, *57*, 211–242. [CrossRef]

32. Fritsche, U.R.; Iriarte, L. Sustainability criteria and indicators for the bio-based economy in Europe: State of discussion and way forward. *Energies* **2014**, *7*, 6825–6836. [CrossRef]

33. Forbord, M.; Vik, J.; Hillring, B.G. Development of local and regional forest based bioenergy in Norway—Supply networks, financial support and political commitment. *Biomass Bioenergy* **2012**, *47*, 164–176. [CrossRef]

34. Soliño, M.; Prada, A.; Vázquez, M.X. Green electricity externalities: Forest biomass in an Atlantic European Region. *Biomass Bioenergy* **2009**, *33*, 407–414. [CrossRef]

35. BIORAISE. Biomass Database. Available online: http://bioraise.ciemat.es/Bioraise/home/main (accessed on 10 March 2020).

36. Verhoeff, F.; Pels, J.R.; Boersma, A.R.; Zwart, R.W.R.; Kiel, J.H.A. *ECN Torrefaction Technology Heading for Demonstration*; ECN (Energy Research Centre of the Netherlands): Petten, The Netherlands, 2011; pp. 1–8.

37. Bergman, P.C.A.; Boersma, A.R.; Zwart, R.W.R.; Kiel, J.H.A. *Torrefaction for Biomass Co-Firing in Existing Coal-Fired Power Stations*; ECN (Energy Research Centre of the Netherlands): Petten, The Netherlands, 2005; pp. 1–71.

38. Paredes-Sánchez, J.P.; López-Ochoa, L.M.; López-González, L.M.; Las-Heras-Casas, J.; Xiberta-Bernat, J. Evolution and perspectives of the bioenergy applications in Spain. *J. Clean. Prod.* **2019**, *213*, 553–568. [CrossRef]

39. Derčan, B.; Lukić, T.; Bubalo-Živković, M.; Durev, B.; Stojsavljević, R.; Pantelić, M. Possibility of efficient utilization of wood waste as a renewable energy resource in Serbia. *Renew. Sustain. Energy Rev.* **2012**, *16*, 1516–1527. [CrossRef]

40. Paredes-Sánchez, J.P.; García-Elcoro, V.E.; Rosillo-Calle, F.; Xiberta-Bernat, J. Assessment of forest bioenergy potential in a coal-producing area in Asturias (Spain) and recommendations for setting up a Biomass Logistic Centre (BLC). *Appl. Energy* **2016**, *171*, 133–141. [CrossRef]

41. Li, H.X.; Edwards, D.J.; Hosseini, M.R.; Costin, G.P. A review on renewable energy transition in Australia: An updated depiction. *J. Clean. Prod.* **2020**, *242*, 118475. [CrossRef]

42. Festel, G.; Würmseher, M.; Rammer, C.; Boles, E.; Bellof, M. Modelling production cost scenarios for biofuels and fossil fuels in Europe. *J. Clean. Prod.* **2014**, *66*, 242–253. [CrossRef]

43. Proskurina, S.; Heinimö, J.; Schipfer, F.; Vakkilainen, E. Biomass for industrial applications: The role of torrefaction. *Renew. Energy* **2017**, *111*, 265–274. [CrossRef]

44. Lee, S.Y.; Sankaran, R.; Chew, K.W.; Tan, C.H.; Krishnamoorthy, R.; Chu, D.T.; Show, P.L. Waste to bioenergy: A review on the recent conversion technologies. *BMC Energy* **2019**, *1*, 1–22. [CrossRef]

45. Moya, R.; Tenorio, C.; Oporto, G. Short rotation wood crops in Latin American: A review on status and potential uses as biofuel. *Energies* **2019**, *12*, 705. [CrossRef]

46. Ubando, A.T.; Felix, C.B.; Chen, W. Biorefineries in circular bioeconomy: A comprehensive review. *Bioresour. Technol.* **2020**, *299*, 122585. [CrossRef] [PubMed]

47. Sahu, S.G.; Chakraborty, N.; Sarkar, P. Coal–biomass co-combustion: An overview. *Renew. Sustain. Energy Rev.* **2014**, *39*, 575–586. [CrossRef]

48. Paredes-Sánchez, J.P.; Míguez, J.L.; Blanco, D.; Rodríguez, M.A.; Collazo, J. Assessment of micro-cogeneration network in European mining areas: A prototype system. *Energy* **2019**, *174*, 350–359. [CrossRef]

49. San Cristóbal, J.R. Multi-criteria decision-making in the selection of a renewable energy project in Spain: The Vikor method. *Renew. Energy* **2011**, *36*, 498–502. [CrossRef]

50. Moiseyev, A.; Solberg, B.; Kallio, A.M.I. Wood biomass use for energy in Europe under different assumptions of coal, gas and CO_2 emission prices and market conditions. *J. For. Econ.* **2013**, *19*, 432–449. [CrossRef]

Article

Non-Intrusive Load Monitoring (NILM) for Energy Disaggregation Using Soft Computing Techniques

Cristina Puente [1,*] , **Rafael Palacios** [2], **Yolanda González-Arechavala** [2] and **Eugenio Francisco Sánchez-Úbeda** [2]

1   Computer Science Department, ICAI School of Engineering, Comillas Pontifical University, 28015 Madrid, Spain
2   Institute for Research in Technology (IIT), ICAI School of Engineering, Comillas Pontifical University, 28015 Madrid, Spain; rafael.palacios@iit.comillas.edu (R.P.); yolanda.gonzalez@iit.comillas.edu (Y.G.-A.); eugenio.sanchez@iit.comillas.edu (E.F.S.-Ú.)
*   Correspondence: cristina.puente@comillas.edu

Received: 31 March 2020; Accepted: 4 June 2020; Published: 16 June 2020

Abstract: Non-intrusive load monitoring (NILM) has become an important subject of study, since it provides benefits to both consumers and utility companies. The analysis of smart meter signals is useful for identifying consumption patterns and user behaviors, in order to make predictions and optimizations to anticipate the use of electrical appliances at home. However, the problem with this kind of analysis rests in how to isolate individual appliances from an aggregated consumption signal. In this work, we propose an unsupervised disaggregation method based on a controlled dataset obtained using smart meters in a standard household. By using soft computing techniques, the proposed methodology can identify the behavior of each of the devices from aggregated consumption records. In the approach developed in this work, it is possible to detect changes in power levels and to build a box model, consisting of a sequence of rectangles of different heights (power) and widths (time), which is highly adaptable to the real-life working conditions of household appliances. The system was developed and tested using data collected at households in France and the UK (UK-domestic appliance-level electricity (DALE) dataset). The proposed analysis method serves as a basis to be applied to large amounts of data collected by distribution companies with smart meters.

Keywords: NILM; disaggregation methods; non-intrusive load monitoring; appliance consumptions; soft computing

---

1. Introduction

Two of the main global problems that we are currently facing are pollution and consumption control. In the Paris COP (United Nations Climate Change Conference 2015), the United Nations agreed to limit global warming to 1.5 degrees by 2100 and, therefore, reducing energy consumption has become a key task for achieving this goal [1]. The fact that 27% of electricity consumption in Europe is attributed to households emphasizes the need to enact regulations promoting suitable and responsible electricity usage. In this sense, monitoring home energy consumption is an important task in order to optimize and reduce electricity usage [2,3]. Therefore, there will be great benefits if behavioral patterns on appliance usage could be automatically detected with an eye to modifying consumer habits [4], with a potential reduction of 12%, depending on the type of feedback that is provided, as depicted in Figure 1.

Thus, electricity consumption has been studied progressively more, since it has benefits for both sides: consumers and the energy companies. With regard to consumers, consumption control reduces their demand for energy [5], as feedback in this area is proven to lead to a reduction in billing of 3% to

12% [6,7]. Further, disaggregation can be used to detect broken appliances [8] and to check if appliances were left on, as proposed by Bidgely [9], by using a smartphone application. With recent and upcoming electricity tariffs, which may change dynamically depending on current demand, users may benefit from a smart system that suggests how to delay or advance the running of certain electrical appliances. At this point, the use of smart meters has been proven to be the cheapest and most effective way to control consumption [3].

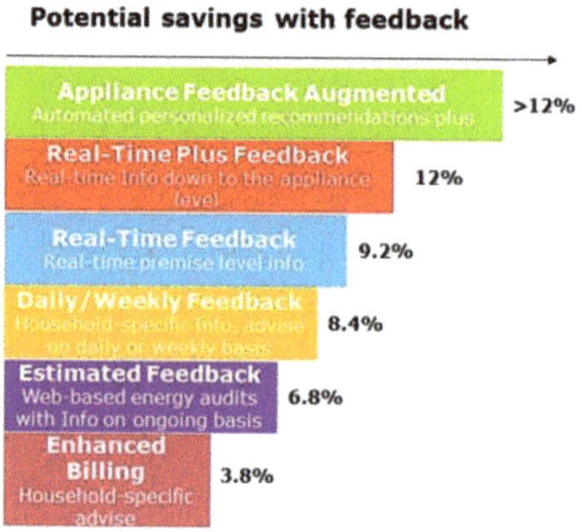

Figure 1. Graphical representation of independent studies analyzed by [3] about residential savings resulting from different types of consumption feedback.

In the case of energy providers, a disaggregated bill can be employed to provide personalized energy saving recommendations [7,10], grid control [11], predictions [12], failure detection [13], and similar statistics.

More than 30% of home consumption is from basic appliances, like the washing machine, refrigerator, and oven. The different behaviors of these appliances make it difficult to detect their patterns based on aggregated consumption. Behavioral patterns may differ in some cases, although, in others, they are relatively close, and could be classified into four general types according to their operational states [14], as shown in Figure 2.

Some appliances are relatively easy to detect, as their behaviors have a characteristic pattern, as shown in Figure 3 [5], for the case of a refrigerator. However, not all appliances have such distinguishable patterns—ranging from kettles to washing machines—which makes their detection a difficult task, as some of their stages behave similarly. At this point, fuzzy clustering plays a role to allow for more than one clustering of classification with different degrees of belonging.

Our proposal identifies consumption patterns by detecting changes in power and building a box model to express consumption as a sequence of elastic bars with different powers and different durations. The system was built and tested using a four-year dataset with standard household consumption data collected via smart meters.

According to this, this paper is organized as follows: Section 2 presents a review of non-intrusive load monitoring (NILM) approaches using different techniques to isolate individual patterns. Section 3 describes the dataset used, along with the most significant fields to be analyzed. Section 4 explains the methodology and techniques used, and Section 5 presents the results obtained, ending with conclusions and future work.

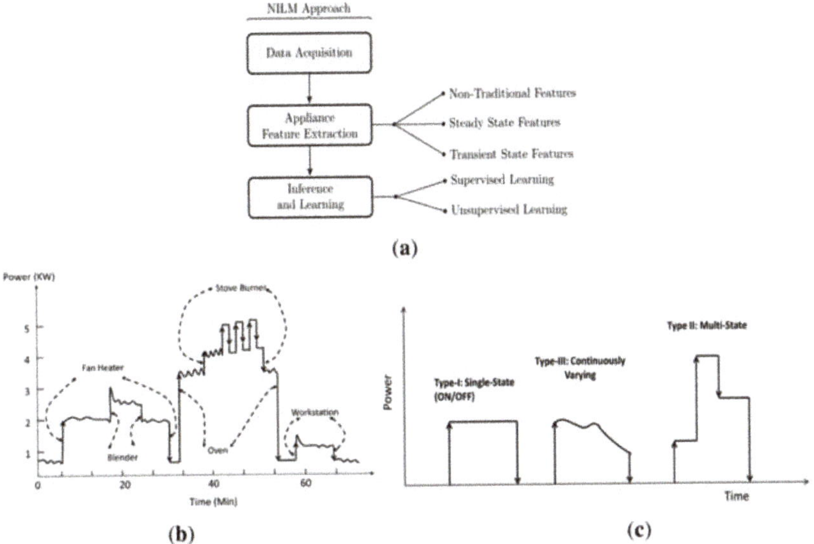

Figure 2. Classification of appliances based on their operation types [14]. (**a**) General framework of NILM approach (**b**) An aggregated load data obtained using single point of measurement; (**c**) Different load types based on their energy consumption pattern.

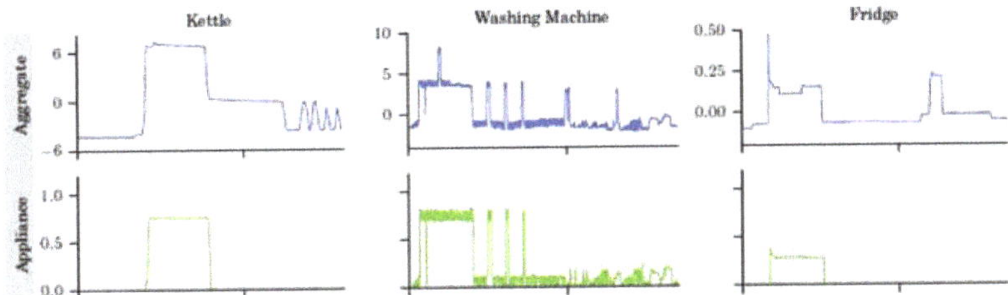

Figure 3. Similarity of stages between appliances. (Reproduced with the permission from [5], 2016).

2. State-Of-The-Art

The study of energy disaggregation is called non-intrusive load monitoring, or NILM, and was patented by George Hart in the 1980s as a basic process for showing the differences that reactive power can provide to distinguish one appliance from another [15], as shown in Figure 4.

Since then, many studies have emerged that have approached the problem from two perspectives:

1. As an optimization problem.
2. As a pattern recognition problem.

Dealing with NILM as an optimization problem is computably unattainable because every appliance has a different set of states. We do not know the consumption of each state, and we do not know the exact number of appliances in a house that are running at the same time. And a further difficulty is that the same appliance often even produces different wave forms, as shown in Figure 5. All these factors, added to the fact that we are handling the aggregated signals of all appliances, end up posing problems with exponential complexity, as mathematically demonstrates Kelly in Reference [5].

In the orientation of NILM as a pattern recognition problem, there are many approaches based on event detection, meaning locating any switch in a signal from a steady state to a new state [16,17]. Algorithms based on event detection, once the event is detected, try to classify the most representative characteristics of a given appliance so as to differentiate and identify them [18]. The standard procedure for this approach is represented in Figure 6.

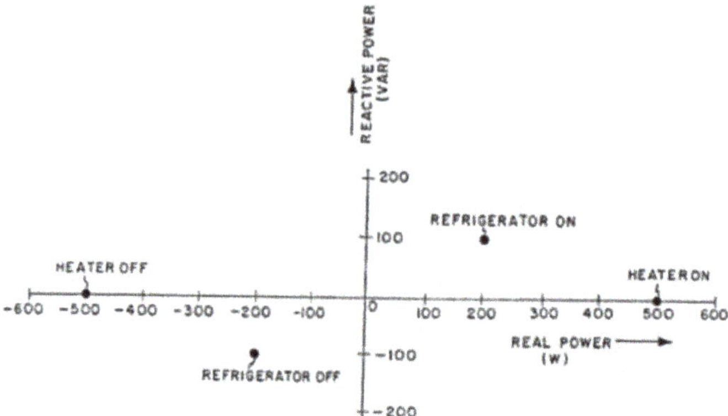

Figure 4. Taken from Hart's US patent number 4,858,141 to map the differences in reactive power to distinguish appliances.

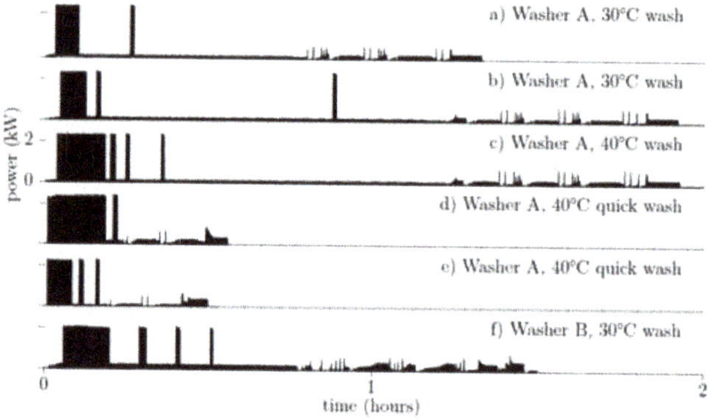

Figure 5. States of a washing machine with different washing programs. (Reproduced with the permission from [5], 2016).

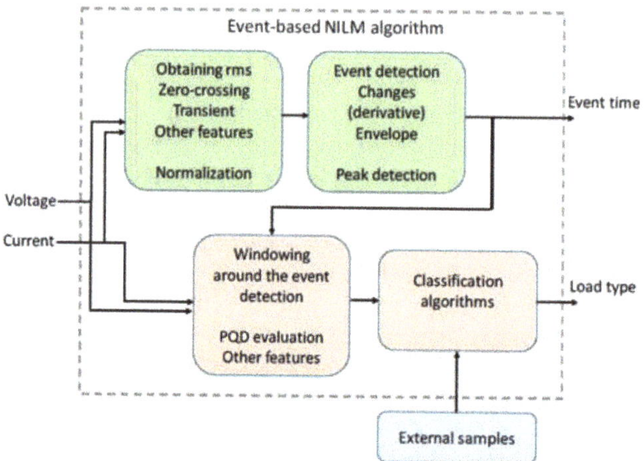

Figure 6. Diagram of event-based non-intrusive load monitoring (NILM) algorithms [18].

There are three main approaches for working with event detection in signals: expert heuristic, probabilistic models, and matched filters [19].

Algorithms based on expert heuristic evaluations try to differentiate appliances by a set of rules with significant variables, such as power variation or power consumption. Probabilistic approaches use models to isolate the concurrence of events. They require training models to adjust variables and create statistical models, as in the case of the generalized likelihood ratio (GLR) method [20]. To compare the heuristic method, see Reference [21]. Yang et al. tested a probabilistic algorithm based on goodness-of-fit (GOF), with results revealing that this method's results were more accurate and had less false positives.

The third type—matched filters—uses patterns that are correlated with the signal waveform to detect the type of appliance. In this case, a large amount of data is required [22].

In this scope, there are many works mixing advance techniques of machine learning, and some other Artificial Intelligence algorithms, as seen in Reference [23], since the application of advanced machine learning techniques as Hidden Markov Models [24–26] until evolved neural networks as BP-ANNs (Back-Propagation-Artificial Neural Networks) in Reference [27] or CNNs (Convolutional Neural Networks) in Reference [28–30].

Our approach is framed in this third group, as it identifies sections in which several appliances can be running simultaneously. These time intervals are adaptable in length, so the problem of devices operating for different lengths of time, with the consequent weakness for pattern correlation, is solved in the proposed approach. Further, the power level identified for each box is discretized using fuzzy clustering techniques, and, consequently, the method can handle the problem of having different sets of devices with similar total power levels.

3. Description of the Dataset

Two different datasets were used in this research. The first dataset comprising electrical consumption in house in France, near Paris, collected by Georges Hebrail. This dataset is available at the Machine Learning Repository of the Center for Machine Learning and Intelligent Systems of the University of California, Irvine [31]. It was used for the development of the proposed algorithm. A second dataset, UK-domestic appliance-level electricity (DALE) 2015 [32], contains aggregated and disaggregated data for 5 houses located in Southern England. It was used for evaluation purposes, and it is described in the Results section.

The Paris dataset consists of a single household's power consumption collected over the course of four years: 2007 through 2010 (precisely from 16 December 2006 17:24:00 to 26 November 2010 21:02:00). A total of 2,075,259 measurements, collected every minute, are included in the dataset. Data was collected at Sceaux (a village located south of Paris, France). This dataset was collected and made public by Georges Hebrail, Senior Researcher, EDF (Électricité de France) R&D.

This dataset contains seven variables (besides date and time), which are:

- global_active_power: The total active power used at the house (kilowatts)
- global_reactive_power: The total reactive power consumed by the household (kilowatts)
- voltage: Average voltage (volts)
- global_intensity: Average current intensity (amps)
- Sub_metering_1: Active energy for kitchen (watt-hours of active energy)
- Sub_metering_2: Active energy for laundry (watt-hours of active energy)
- Sub_metering_3: Active energy for climate control systems (watt-hours of active energy)

Sub_metering_1 is the kitchen, primarily a dishwasher, electric oven, and a microwave oven (hot plates are not electric, but gas powered).

Sub_metering_2 is for the laundry room, containing a washing machine, a tumble dryer, refrigerator, and a light.

Sub_metering_3 is for the heating system, containing a water heater, and an air-conditioning unit.

There is some electrical equipment that is not connected to any of the three sub-meters but directly to the global meter (see Figure 7). Therefore, the sum of Sub_metering_1, Sub_metering_2, and Sub_metering_3 (converted from watt-hours to kilowatts) does not equal to global_active_power. Nevertheless, the objective of this work is to identify different machines and detect when they are in use, so, for this purpose, we started working with the three Sub_metering signals and then demonstrated the approach for the sum of these three signals.

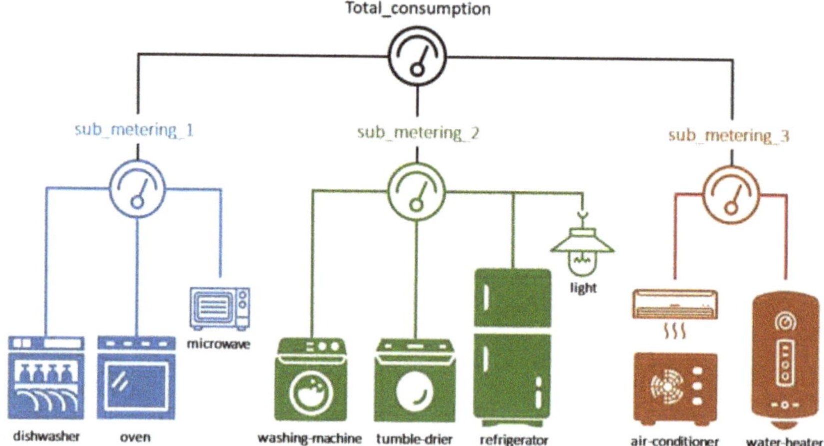

Figure 7. Schematic of electrical appliances and smart meter distribution.

Some preprocessing of the data was necessary to convert energy units to power units and to smooth the values using a filter. Sub_metering data was stored in energy units (watt-hours) every minute. But it is more common to use power units, meaning the average power during the time windows, in this case one minute. Therefore, the energy values of each Sub_metering had to be multiplied by 60 to obtain the average power during every minute (in watts).

Filtering was also very convenient because the power of a small refrigerator is about 100 W, but, in energy per minute, that is only 1.67 Wh. Since the values of the smart meters used to collect the data can only be integers, the values alternate between 1 Wh and 2 Wh. Hence, a Gaussian-weighted moving average filter of size 7 was applied to the data to make it less noisy and more realistic.

4. Data Overview

The Paris dataset was previously analyzed from the time series point of view [33], although standard data series techniques cannot detect the activation of different appliances because they do not show seasonality or fixed-time patterns. As shown in Figure 8, the power profile of Sub_metering_2 is very predictable because it has fixed-time running/waiting cycles. This graph shows three days of data in which practically the only appliance running was the refrigerator. Only the 12th of September shows high-power activity from the washing machine, which overlaps the refrigerator's regular activity.

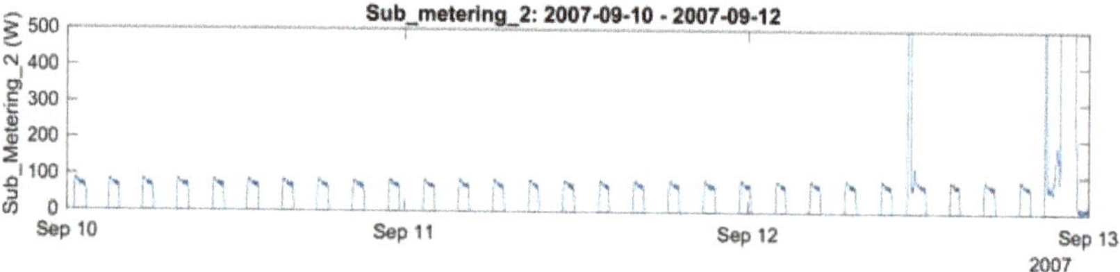

Figure 8. Sub_metering_2 power, during 3 days in September 2007.

Therefore, appliances, like refrigerators, are very predictable, and modeling by means of a time series model is feasible. There are some differences in the period of the signal, which may depend on the thermostat setting for the room's temperature, although neither of them changes very often. So, a model that implements some adaptation and forgetting factors could cope with the signal type without any trouble. As an example, Figure 9 shows the power of Sub_metering_2 in two time intervals in which no appliances were operating other than the refrigerator. The first graph corresponds to 28 June 2010—the middle of the summer—and the refrigerator starts with a period of nearly 2 h, while the second graph corresponds to 17 December 2009—winter—and the period is longer than 3 h.

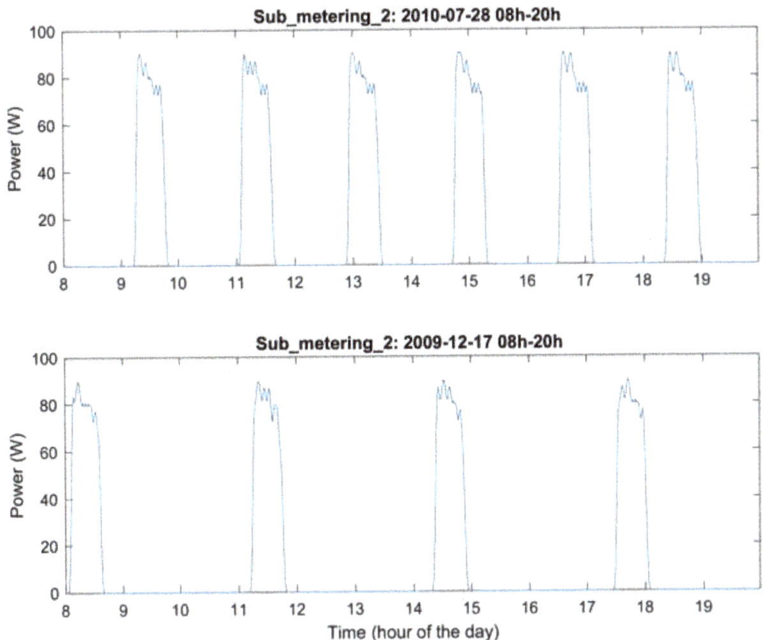

Figure 9. Profiles of the refrigerator in summer and winter.

Conversely, the water-heater basically starts when hot water is used in the house. Figure 10 shows the power profile of Sub_metering_3, which includes the water heater, during the four Mondays in the month of September 2007. There is a power step reaching about 1kW every morning, which is probably triggered by using the shower. The exact time of this event is not always the same (7:05 a.m. on 3 September, 6:15 a.m. on 10 September, 7:24 a.m. on 17 September, and 6:29 a.m. on 24 September). In addition, the length of time that the water heater runs was not constant, where the variation is probably due to the amount of water used. These parameters (start time and elapsed time) depend on user behavior and cannot be predicted with time-series analysis techniques. Even when selecting the same day of the week, as in Figure 10, which should be the most similar to each other, the power profiles are completely different.

The proposed methodology to identify which electrical appliances are installed and their usage patterns involves the use of several techniques. Firstly, regression trees [34] are used to determine the instants of power change, as well the different consumption levels in the house. This step also lets consumption boxes with variable time lengths be detected for each power level. Secondly, clustering techniques are applied to the power levels, to minimize the effects of noisy power measurements, as well as to ascertain which power levels are actually relevant.

This approach could be implemented massively at the level of the electricity utility by using data stream models, such as the one proposed in Reference [35].

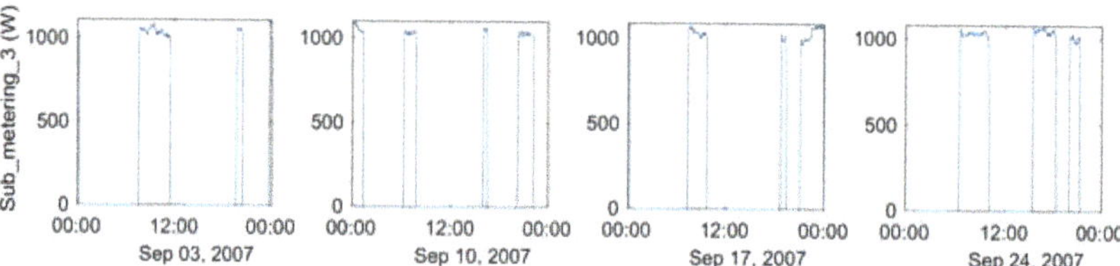

Figure 10. Sub_metering_3 power, during the four Mondays in September 2007.

5. Fuzzy Clustering

In an imprecise environment, like the one we have in our systems, soft computing techniques have emerged to model imprecise scenarios [36,37]. Clustering techniques are very popular as supervised methods that are used to classify information according to a set of properties. In NILM problems, clustering algorithms have been used to isolate patterns in several groups, combining them with other procedures to obtain better results in most cases than when using traditional clustering. In Reference [38], Liu et al. used fuzzy clustering techniques to create a set of general models that could better detect appliances within a household. Wang et al. used fuzzy clustering, along with Hidden Markov models [26], to retrieve single energy consumption based on the typical consumption pattern. Lin [23,39] proposed a hybrid system using fuzzy clustering, along with neural networks, leading to the identification of household appliances claiming an accuracy of 95%. In Reference [40], Kamat used fuzzy logic applied to pattern recognition, in particular to detect the period of operation of a given device and thus calculate the energy consumed by that particular device.

The method that we propose is focused on the analysis of the aggregated consumption signal. In the case of several appliances operating at the same time, the consumption profiles overlap, making pattern recognition technique difficult to apply to the aggregated signal.

We use fuzzy logic and regression trees to create a box model to model changes in power and to discretize power level, allowing to differentiate devices even if the operate simultaneously. Neural methods, as Reference [39,41,42], need intensive training to adjust the neural network. In our case, being an unsupervised method, we do not need large amounts of data to train our method and obtain good results, as shown in the following sections.

In our case, by segmenting electrical information, the appliances are grouped based on their "distance," understanding this distance from a mathematical view as their closeness to each other to define two objects' similarity. According to these groupings, we have two clustering types:

- Hard clustering: the traditional version, where objects can belong to just one group.
- Fuzzy clustering: this technique uses fuzzy logic [43], allowing the same object to be in more than one group. The difference is the degree of membership or the extent to which each object belongs to that cluster [44]. This approach is closer to real-life problems. In our case, we can isolate two "similar" signals from the aggregated signal to detect the device to which it belongs.

Therefore, we use an objective function to obtain the optimal number of partitions that will let us apply non-linear optimization algorithms to find a local minimum.

We have to define the number of clusters according to these three conditions, with c being the number of clusters and N the number of items:

$$\mu_{ij} \in [0,1], \ 1 \leq i \leq c, \ 1 \leq j \leq N$$
$$\sum_{i=1}^{c} \mu_{ij} = 1, \ 1 \leq j \leq N \tag{1}$$
$$0 < \sum_{j=1}^{N} \mu_{ij} < N, \ 1 \leq i \leq c.$$

With this scenario, we define our fuzzy space as:

$$F_c = \left\{ U \in \mathbb{R}^{cxN} \middle| \mu_{ij} \in [0,1], \forall i,k; \sum_{i=1}^{c} \mu_{ij} = 1, \forall k; 0 < \sum_{i=1}^{c} \mu_{ij} < N, \forall i \right\}. \tag{2}$$

Fuzzy clustering c-means is based on the optimization of fuzzy partitions [45,46], with U being the membership matrix $[\mu_{ij}] \in F_c$, and $V = [v_1, v_2, \ldots, v_c]$ being the vectors characterizing the centers of these groupings, for which we want to minimize our function.

$$J(Z, U, V) = \sum_{i=1}^{c} \sum_{j=1}^{N} (\mu_{ij})^m \|z_j - v_i\|_A^2. \tag{3}$$

The value of the cost function $J(Z, U, V)$ can be interpreted as a measure of the deviation between points v_i and centers z_j.

The minimization of this function leads to a non-linear optimization problem solved by the Picard iterative process. The restriction of membership values, μ_{ij}, is imposed by Lagrange multipliers.

$$J(Z, U, V) = \sum_{i=1}^{c} \sum_{j=1}^{N} (\mu_{ij})^m \|z_j - v_i\|_A^2 + \sum_{j=1}^{N} \lambda_j \sum_{i=1}^{c} (\mu_{ij} - 1). \tag{4}$$

We can demonstrate that, to minimize the function, it is necessary that:

$$\mu_{ij} = \frac{1}{\sum_{k=1}^{c} (D_{ijA}/D_{kjA})^{2/(m-1)}}, \ 1 \le i \le c, \ 1 \le j \le N$$

$$v_i = \frac{\sum_{j=1}^{N} (\mu_{ij})^m z_j}{\sum_{j=1}^{N} (\mu_{ij})^m}, \ 1 \le i \le c. \tag{5}$$

Therefore, we need some other parameters for the algorithm, such as the number of clusters, which is one of the most relevant due to having a great impact on segmentation. The number of clusters is obtained through the fuzzy partition coefficient (FPC), which provides how well our data are explained by this grouping, that is, that membership to each one of our data segments is—in general—strong and not fuzzy. The fuzziness parameter, m, which affects fuzziness in the segmentation, is completely fuzzy if it approaches ∞ and hard as it approaches 1. In our case, we set a value of ($m = 2$), as a standard value for these types of problems, which is widely used in the bibliography. As termination criteria, we established X number of iterations and the distance matrix. This is because the calculation of distance implies establishing the scalar product matrix. The natural choice is the identity matrix ($A = I$), but a widespread distance matrix is the inverse of the covariance matrix of the data, leading to the Mahalanobis standard.

$$A = R^{-1}, \ R = \frac{1}{N} \sum_{i=1}^{N} (z_i - \bar{z})(z_i - \bar{z})^T. \tag{6}$$

The norm used affects to the segmentation criteria, changing the measure of dissimilarity. The Mahalanobis distance leads to hyperellipsoid groupings on the axes, given by the covariances between variables.

In the bibliography, there are several modifications of this algorithm related to use an adaptative distance measure [47,48] and relaxing the condition on probability of belonging to each segment. According to these parameters, we checked the Euclidean norm, Mahalanobis, and Gustafson-Kessel algorithm.

The Gustafson-Kessel algorithm expanded the adaptive distance to locate different groupings with distinct geometrical forms. Each segment has its own distance provided by the equation:

$$D_{ijA_i}^2 = (z_j - v_i)^T A_i (z_j - v_i). \tag{7}$$

The matrices A_i become variables that are optimized within the functional J. The only restriction is that the determinant must be positive, $(|A_i| = \rho_i,\ \rho_i > 0,\ \forall i$. Optimizing by using the Lagrange multipliers method, we obtain that the distance matrices must fulfill this equation:

$$A_i = [\rho_i \det(F_i)]^{1/m} F_i^{-1}, \tag{8}$$

where F_i is the fuzzy covariance matrix of each one of the segments.

$$F_i = \frac{\sum_{j=1}^{N} \left(\mu_{ij}\right)^m \left(z_j - v_i\right)\left(z_j - v_i\right)^T}{\sum_{j=1}^{N} \left(\mu_{ij}\right)^m}. \tag{9}$$

We checked several measures to verify which ones fit the best to segment our datasets [49].

6. Description of the Analysis Procedure

The proposed analysis approach involves several steps that are described in this section. Some of these algorithms are shown using the signal obtained by one of the smart meters, but this is just for clarification purposes, since the whole approach has been designed to be implemented on the signal measured by a single smart meter that obtains the global household power consumption. If we could expect the signals of several smart meters to be available on a regular basis in a standard household, there would be a separation of appliances that would facilitate the analysis greatly. For instance, this would make it possible, and very effective, to obtain the appliances' typical operating patterns, such as the most-used dishwasher and washing machine cycles. Then, by applying pattern recognition techniques, it would be straightforward to detect when and how the appliances are used. However, this approach becomes very problematic when trying to analyze one global signal for the household—the sum of all the Sub_metering signals—because the power profiles of all the appliances in the house become all mixed up in the single power signal.

Nevertheless, the proposed approach identifies instants of significant power changes, power levels, and length of conditions, therefore defining a sequence of boxes of different heights and widths with which to model global power consumption. Using this box-based model, it is possible to identify which appliances are being used and when, which will enable a higher-level analysis of weekly or seasonal usage patterns. This higher-level information about usage patterns would be extremely useful for determining if small changes in schedules and habits can benefit the power network and harvest savings for the user.

6.1. Detection of Power Changes

A very simple way to detect changes in power is to work with the derivative of the power signal, although this method will be subject to many errors, due to the noise expected in the signal. Another widely-used method, and more robust, is the MATLAB function findchangepts, which is based on the algorithm described by Killick et al. [50]. However, using this method, we found an average of 522 change points per week in the dataset of just one Sub_metering. This is much higher than expected, compared to a naked-eye detection of the signal, especially knowing that the only electrical appliance running was the hot water heater, in cycles with a fairly constant operational power level.

The proposed approach is to use regression trees [34] to detect power changes. The tree can make a decision based on the values of the signal to determine whether or not level changes are significant to the problem. In contrast to the decision tree, created with algorithms, such as ID3, the training process of regression trees is unsupervised. Hence, it is not necessary to manually label which changes are relevant and which ones are noise. The regression tree can be automatically applied to any household without prior knowledge of the appliances installed and without any manual pre-analysis and annotation of the signals.

Trees must be pruned to avoid overfitting, to make them more generic, and to yield better overall results. The model developed in this way is very accurate and could clearly detect all 25 water-heater cycles during one week in September 2007 of data analyzing smart meter number 3, in which the air conditioning unit was not available yet. Figure 11 shows the actual power of Sub_metering_3 in blue and the prediction of the model in red. The second graph shows the prediction error, which only has spike values during transients of power.

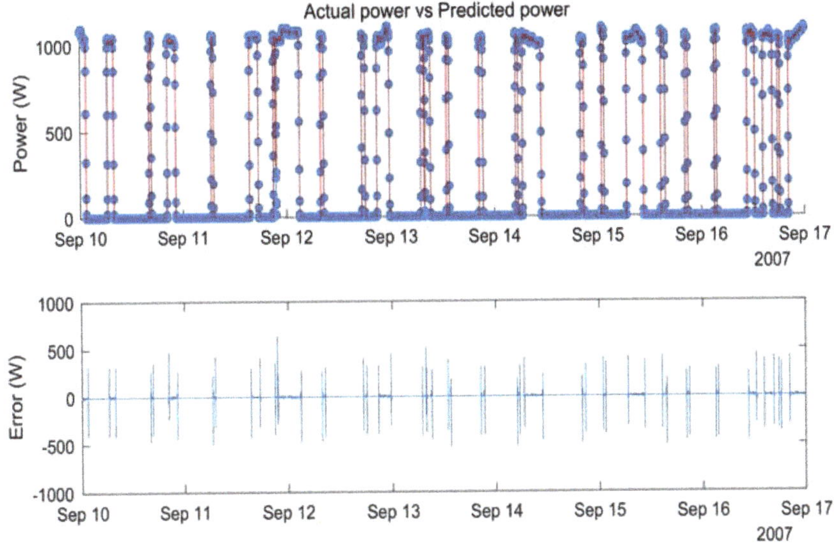

Figure 11. Results of the regression tree for one week of data.

6.2. Power Levels

The regression tree model, described in the previous section, is able to predict the instant of power changes, along with the power level. However, the power levels in a global power signal are linear combinations of the power levels of the appliances installed in the house. Consequently, in order to generate boxes with a meaningful height, which will not be affected by signal noise, some data must be analyzed to determine what the typical power levels are in a given household.

By doing clustering analysis, we can obtain the "standard" levels of power of a given household. As depicted in Figure 12 there are many standard levels that appear a significant number of times. The application of clustering techniques at this point helps to detect power levels without the hassle of small power changes due to the normal operation of any electrical machine.

As described in previous sections of this paper, in current applications, the use of fuzzy clustering is more suitable because the subsequent analysis of which appliance set was operating at a given time is easier if a given level can belong to several clusters and not just one. In this way, the number of clusters is reduced, and the impact of noise is even more mitigated.

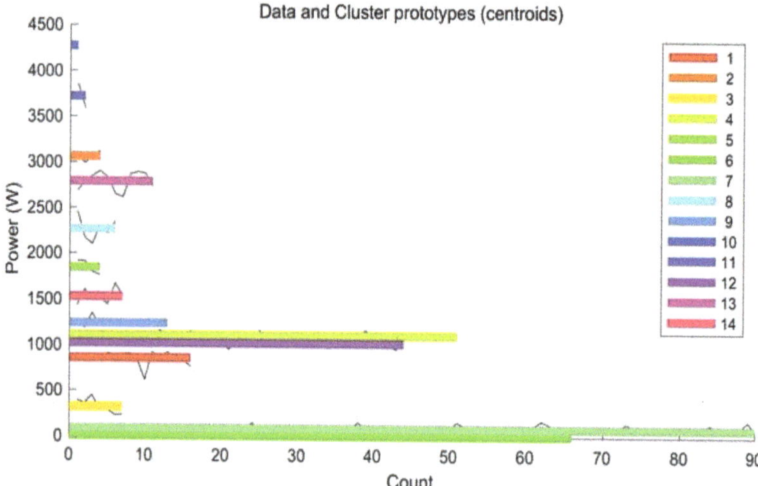

Figure 12. Results of the clustering of power levels of the global consumption signal.

6.3. Box Model

The final step in the modeling and, hence, understanding the shape of the power signal, is to produce the box model. In the proposed approach, the power signal is symbolized by a temporary sequence of rectangles of different heights and widths, which we, the authors, named the Box Model. This type of signal representation can handle the problems, described in the Data Analysis section, related to the length and temporary spacing of some cycles, such as refrigerator operation, which is different in summer and winter, or the water heater, which primarily depends on the amount of water used.

As a result of the Box Model, the first boxes of a real signal analyzed by this approach are shown in Table 1. The values that completely define a box are the starting point in minutes, the height of the box in watts, and the width of the box in minutes. Each box represents a state of the system, and, whenever system conditions change, a new box is created.

Table 1. Example of Box Model values.

FROM (min)	TO (min)	HEIGHT (W)	WIDTH (min)
1	30	1085	30
31	45	1139	15
46	70	1111	25
71	80	1072	10
81	90	108	10
91	176	0.4	86
177	212	75.5	36
213	223	59.8	11

6.4. Detection of Appliances

Each box corresponds to a state in the system, since any significant change in the power will produce a new box. These boxes are classified, based on the mean power level, applying clustering techniques. Using classic clustering, such as K-means, each box will be assigned to the closest cluster, so the appliances that are active for a given box are those represented in the cluster. The centroid of each cluster is a power level related with the electrical appliances that are in use simultaneously. One may think of the different cluster centroids as linear combinations of the operational power level of the appliances.

In contrast with classic clustering, using fuzzy clustering, one obtains, for a given box, the membership degree of that box belonging to each of cluster. Therefore, the classification of each box does not yield to a single answer, and several combinations of appliances could be considered. In general, just one of the clusters attains a high and distinctive degree, hence behaving as classic clustering. However, in some situations, the power level of the box may well represent to possible configurations of appliances. In the upcoming Results section, an example of the potential of fuzzy clustering is presented.

7. Experimental Results

In order to show the effectiveness of the Box Model proposed in this paper, all four years of data of the Paris dataset where analyzed. Figure 13 shows the number of boxes created for every year and every month, which is an indication of the activity in the house, since each change in the consumption level creates a new box. In 2010, the graph does not include November and December because data collection runs until mid-November of that year. In 2007, the last week of July and the first 3 weeks of August there was very little activity because the house owners were probably on vacation; so, even though June through September are the hottest months of the year in Paris and Air Conditioning activity could be expected, the overall activity in 2007 is smaller. In 2008, August was the period of summer vacation and the level of activity was also smaller that in July of September.

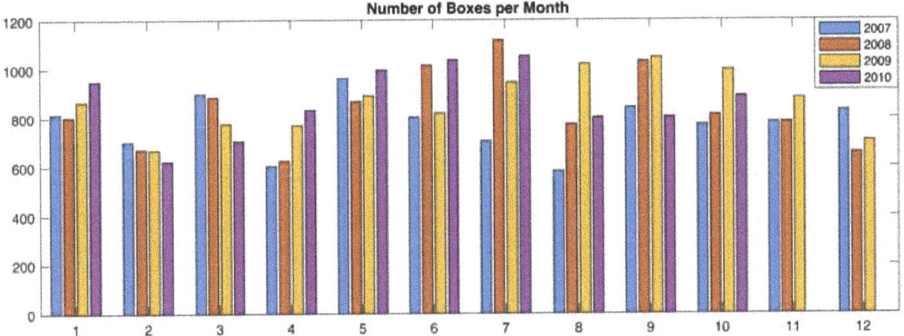

Figure 13. Results of the Box Model for all four years of data, sorted by month.

As an example of the effectiveness of the proposed methodology for detecting the use of electrical appliances, Figure 14 shows a challenging scenario in which several appliances are operating simultaneously, thus making the resulting total household power signal difficult to analyze. Since the dataset used for this work uses smart meters for different sections of the house, it is possible to understand the shape of the total power (subplot 4) by looking at the decomposed signals in subplots 1 through 3.

The first graph in Figure 14 shows Sub_metering_1, which includes a consumption event that starts at 10:42 a.m. This event very likely corresponds to the dishwasher. The profile has a total duration of 80 min and is characterized by a flat power level of about 75 W, with two heating cycles 15 min long and with 2200 W of power.

The second graph shows Sub_metering_2, with very regular refrigerator cycles. This profile corresponds to an old Liebherr refrigerator that is operating every 2 h and 40 min for a period of 45 to 50 min, with a power level between 65 and 85 W. This is the most regular Sub_metering signal in the results represented in the figure.

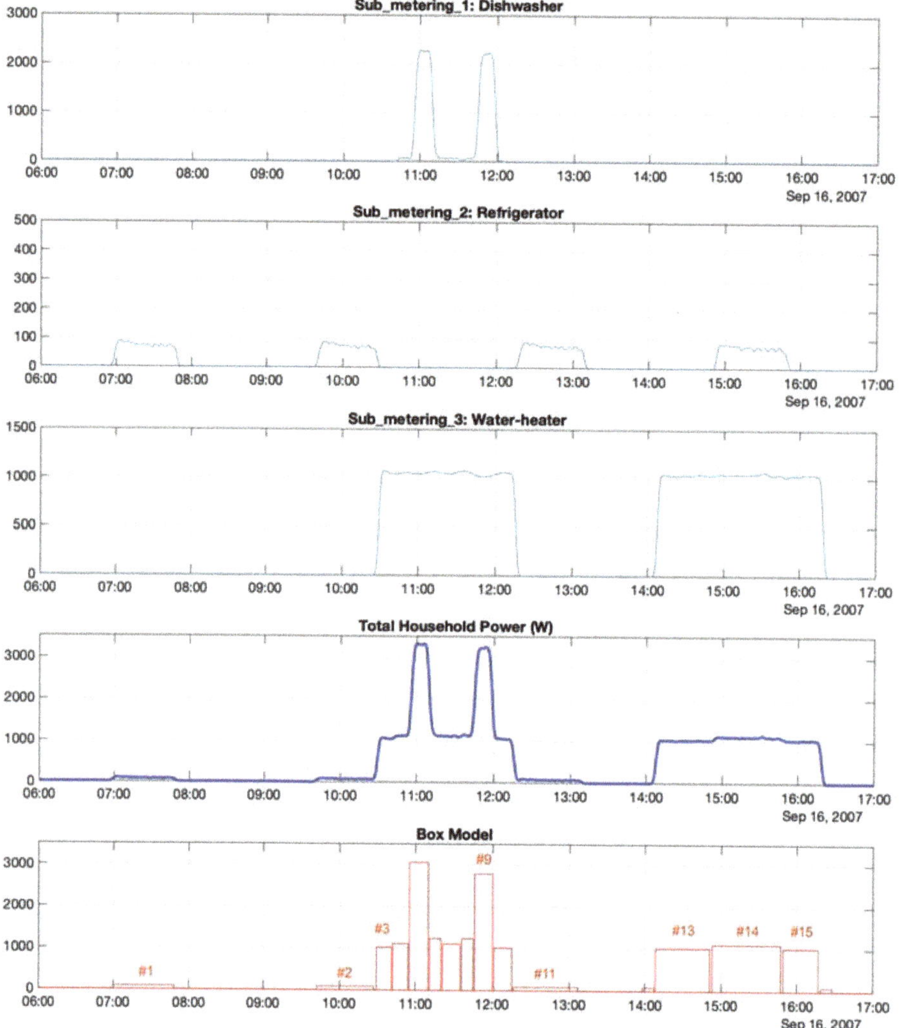

Figure 14. Box Model example.

The third graph shows Sub_metering_3, which measures the consumption of a water heater and an air conditioner. In Figure 14, the water heater operates twice with a fairly constant power level of 1050 W. It starts for the first time at 10:30 a.m. for 1 h and 45 min and the second time starting at 2:10 p.m. with a duration of 2 h 10 min. The figure reveals that the first cycle of the water heater overlaps with the dishwasher, and the second cycle overlaps with the refrigerator, making this example interesting and challenging.

The fourth graph in Figure 14 shows the total power, which is the sum of the three previous smart meter signals. This is the signal that was analyzed using the proposed procedure in order to create the Box Model represented in the fifth graph. The first three graphs are shown in the figure to explain the results obtained by the method.

In the resulting Box Model (fifth graph), the first couple small identical bars, as well as bar #11, are refrigerator cycles and could be easily identified by any data analysis technique. Then, there is bar

#3, which is the water heater, where this profile overlaps the dishwasher in Sub_metering_1. Therefore, after 22 min at 1020 W of power, the algorithm creates a new box (bar #4) with a power level of 1110 W, representing the water heater plus the first part of dishwasher profile. Bars #5 and #9 correspond to the water heater plus high-power (heating) cycles of the dishwasher running.

Another interesting event in this graph is the second cycle of the water heater, which starts at 2:10 a.m. and overlaps with a refrigerator cycle at 2:55 p.m. The algorithm can detect the starting point (bar #13), after a small transient bar #12, which should be removed in future versions of the algorithm. Then, at 2:55 p.m., a change in power is triggered by the refrigerator cycle. The proposed method is able to detect the power change and identify a new power level of 1110 W, which was previously selected as one of the important power levels by the clustering module. Consequently, the method produces three boxes, two corresponding to the water heater alone (bar #13 and bar #15), with a power level of 1020 W, and one for the water heater and the refrigerator (bar #14), with a power level of 1110 W.

For further comparison, the algorithm was run on UK-DALE dataset, that has been widely used in the literature [30,41,42,51–56]. Data was collected from November 2012 to May 2015, with sampling periods of 1s for aggregated and 6s for disaggregated signals. However, not all the houses cover the full range of collection time, and, in fact, there is no period of time in which data was collected from the 5 houses simultaneously. For some houses the disaggregated signals comprised several electrical appliances, so it was decided to use house 2 with 20 data channels and very fine disaggregation. The advantage of house 2 is that all appliances are well identified in separated channels of data so a ground truth for evaluation purposes can be easily generated. This dataset was preprocessed to obtain data samples every minute and selecting similar appliances as in the case of the French dataset. House 2 has interruptions in data collection at different times in different channels, so the month of July 2013 was selected as the best period of time in terms of data quality showing minimal events of missing data in 8 of the 20 channels. Signals were pre-processed to adjust the sampling period to 1 min, as in the other dataset.

Table 2 shows the average power, maximum power and total energy of each channel. Only those appliances highlighted in the table have a significant impact on the aggregated power, since other devices are less relevant for low power or marginal use. For example, small electronic devices are not interesting if operating all the time, such as the router that was only restarted 4 times in the month (was operating 99.96% of the time). The microwave is used every day but very short periods of time, mostly for less than 2 min. Finally, toaster and cooker are demanding in power, but the toaster was only used once, and the cooker was never used. A similar type of selection to focus on the relevant appliances was also done in Reference [51,55].

For each individual channel it is necessary to determine if the appliances were working or not, resulting in a vector of 1s and 0s that will be as the ground truth for evaluating the results. These binary vectors were obtained applying thresholds for each signal.

The dataset was evaluated using a system of 8 fuzzy clusters. A total of 2048 boxes were created automatically for the UK-DALE dataset, which is higher than in the case of the Paris dataset. It could be expected that in a more recent dataset the electrical appliances should be more efficient; nevertheless, the fridge in the UK dataset is less efficient that the refrigerator in the Paris dataset and produces more cycles. The results are typically evaluated in the literature using F1_score, which is defined as:

$$F1\_score = 2\frac{Precision \cdot Recall}{Precision + Recall}, \tag{10}$$

with Precision being the number of true positive divided by predicted positive (how many predictions are correct) and the Recall being the number of true positive divided by the condition positive (how many expected events have been correctly found).

The final results are presented on Table 3 separated by appliances. It can be seen that Kettle and Washing Machine are harder to predict, but the fridge is detected remarkably well.

Table 2. Summary of appliances in UK-domestic appliance-level electricity (DALE) house 2 during July 2013.

Channel	Average (W)	Maximum (W)	Total Energy (kWh)
2 laptop	8.1	59	6.0
3 monitor	19.9	79	14.8
4 speakers	5.8	11	4.3
5 server	14.0	18	10.4
6 router	6.0	7	4.5
7 server_hdd	1.0	1	0.7
8 kettle	**19.5**	**2995**	**14.5**
9 rice_cooker	3.6	414	2.7
10 running_machine	2.4	321	1.8
11 laptop2	4.5	69	3.3
12 washing_machine	**10.7**	**2221**	**8.0**
13 dish_washer	**37.3**	**2064**	**27.7**
14 fridge	**53.0**	**117**	**39.5**
15 microwave	5.3	1330	3.9
16 toaster	0.6	896	0.4
17 playstation	1.0	32	0.7
18 modem	9.0	10	6.7
19 cooker	0.2	392	0.1
1 aggregate	**201.8**	**5105**	**150.1**

Table 3. Results of the algorithm evaluated on UK-DALE. TP = True positive, FP = False positive, FN = False negative, TN = True negative.

Appliance	TP	FP	FN	TN	Prediction	Recall	F1_Score
Kettle	151	108	169	44211	0.58	0.47	0.52
Washing_machine	95	126	145	44273	0.43	0.40	0.41
Dish_washer	723	266	77	43573	0.73	0.90	0.81
Fridge	23751	730	863	19295	0.97	0.96	0.97
AVERAGE					0.68	0.68	0.68

These results are consistent with previous works found in the literature. The results in Reference [55] for houses 1, 2, and 5 together, show a slightly better overall F1_score of 0.77 in the best combination of methods, compared to 0.68. However, in Reference [51] the results for the proposed H-ELM method are very similar: 0.67 in average F1-score, with a minimum average F1 of 0.47 for the washing machine and a maximum of 0.89 for the fridge. It is very interesting that Reference [56] obtains the same average results of 0.63 but points out that using active and reactive power (if available) the results may improve to 0.70.

In regard to the fuzzy clustering approach presented in this paper, Figure 15 presents the case of a Kettle event that is not correctly classified with classic clustering. In these graphs, the black dotted lines represent the centroids of the clusters associated to the dish washer (cluster #7) and to the kettle (cluster #4). Due to sampling effects, the mean power of the box in the left graph is only 2253 W high, which is lower than the typical power near 2500 W of cluster #4. As a consequence of the low power, this event would be classified as a dish washer because it is closer to cluster #7 than to cluster #4. However, with fuzzy clustering, this event was assigned a 0.496 membership degree to dish washer cluster but also a 0.432 membership degree to kettle cluster. Given that a dish washer always runs longer cycles, in this case, the system correctly selected kettle for the event. The rule that was introduced takes into account a low membership (less than 0.5 for cluster #7) and a narrow box (less than 4 min), and, in that case, the system takes the second probable cluster (cluster #4). Once this rule was introduced and run for the full dataset, the number of false positive in dish washer was reduced from 274 to 266, and the number of false negative in kettle was reduced from 175 to 169.

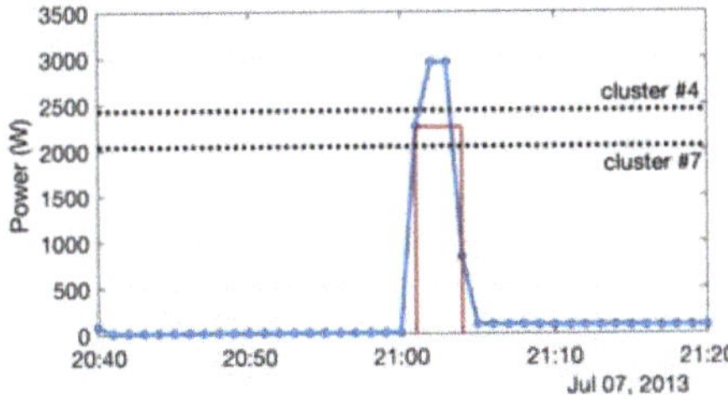
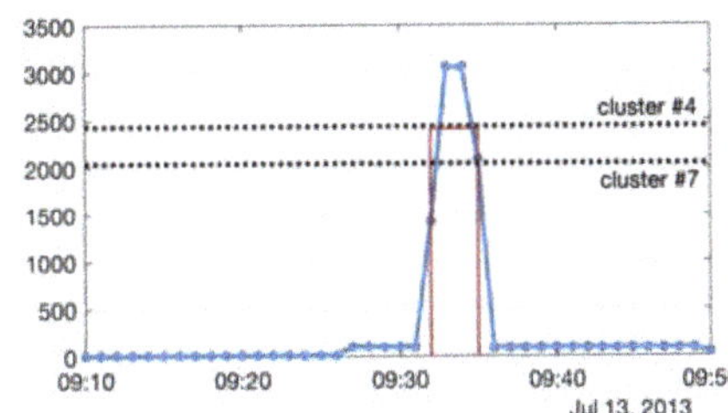

Figure 15. Explanation of the advantage of using fuzzy clustering. Left: Kettle event miss-classified with classic clustering. Right: typical Kettle event (for comparison). Blue line shows the actual power, red line is the result of the box model, and black dotted lines represent the centroids of cluster #4 and #7.

8. Conclusions and Future Work

Monitoring power load makes it possible to obtain usage patterns of electrical appliances, as well as real home energy consumptions. Understanding these data, small behavioral changes could be introduced that could significantly reduce costs and the environmental impact of electricity consumption. However, analyzing aggregated power consumption data is a challenging problem, especially if a previous interaction with individual electrical appliances is not possible. This field is called non-intrusive load monitoring and has been studied for years.

The work presented in this paper involves a methodology to analyze data collected with smart meters, which are currently deployed in many countries or are being installed at this time. The proposed method involves techniques for detecting changes in power based on regression trees, the selection of standard power levels of a household based on fuzzy clustering, and the creation of a Box Model to describe the aggregated power load measured by the smart meter. The system has been developed and tested using data collected in standard households by several smart meters in France and the UK. The results show that the proposed method is able to detect the usage of appliances, even in difficult situations in which several appliances overlap in time. The application of fuzzy clustering solves several cases in which classic clustering miss-classifies the events.

For future works, as fuzzy clustering allows to reduce the number of false positive and false negative, we plan to continue this analysis by the addition of new appliances.

Author Contributions: Conceptualization, E.F.S.-Ú., R.P., C.P. and Y.G.-A.; Methodology, Y.G.-A.; Software, E.F.S.-Ú., R.P. and C.P.; Validation, R.P. and C.P.; Formal analysis, Y.G.-A.; Investigation, Y.G.-A.; Resources, C.P.; Data curation, E.F.S.-Ú., R.P.; Writing—Original draft preparation, C.P. and R.P.; Writing—Review and editing, C.P. and R.P. All authors have read and agreed to the published version of the manuscript.

Funding: This research received no external funding.

Conflicts of Interest: The authors declare no conflict of interest.

References

1. The Paris Agreement|UNFCCC. Available online: https://unfccc.int/process-and-meetings/the-paris-agreement/the-paris-agreement (accessed on 30 March 2020).

2. Wagner, L.; Ross, I.; Foster, J.; Hankamer, B. Trading off global fuel supply, CO2 emissions and sustainable development. *PLoS ONE* **2016**, *11*. [CrossRef] [PubMed]

3. Carrie Armel, K.; Gupta, A.; Shrimali, G.; Albert, A. Is disaggregation the holy grail of energy efficiency? The case of electricity. *Energy Policy* **2013**, *52*, 213–234. [CrossRef]

4. Darby, S.; Liddell, C.; Hills, D.; Drabble, D. *Smart Metering Early Learning Project: Synthesis Report*; DECC (Department of Energy & Climate Change): London, UK, 2015.

5. Kelly, D.G. *Disaggregation of Domestic Smart Meter Energy Data*; London University: London, UK, 2016.

6. Davis, A.L.; Krishnamurti, T.; Fischhoff, B.; Bruine de Bruin, W. Setting a standard for electricity pilot studies. *Energy Policy* **2013**, *62*, 401–409. [CrossRef]

7. Fischer, J.E.; Ramchurn, S.D.; Osborne, M.A.; Parson, O.; Huynh, T.D.; Alam, M.; Pantidi, N.; Moran, S.; Bachour, K.; Reece, S.; et al. Recommending energy tariffs and load shifting based on smart household usage profiling. In *IUI: Proceedings of the International Conference on Intelligent User Interfaces*; Association for Computing Machinery: New York, NY, USA, 2013; pp. 383–394.

8. Chang, H.H.; Lee, M.C.; Chen, N.; Chien, C.L.; Lee, W.J. Feature extraction based hellinger distance algorithm for non-intrusive aging load identification in residential buildings. In Proceedings of the 2015 IEEE Industry Applications Society Annual Meeting, Addison, TX, USA, 18–22 October 2015.

9. Home Energy Reports—Bidgely. Available online: https://www.bidgely.com/bidgely_home-energy-reports/ (accessed on 31 March 2020).

10. Makriyiannis, M.; Lung, T.; Craven, R.; Toni, F.; Kelly, J. Smarter electricity and argumentation theory. In Proceedings of the Smart Innovation, Systems and Technologies; Springer Science and Business Media Deutschland GmbH: Berlin, Germany, 2016; Volume 46, pp. 79–95.

11. Kavgic, M.; Mavrogianni, A.; Mumovic, D.; Summerfield, A.; Stevanovic, Z.; Djurovic-Petrovic, M. A review of bottom-up building stock models for energy consumption in the residential sector. *Build. Environ.* **2010**, *45*, 1683–1697. [CrossRef]

12. Torriti, J. A review of time use models of residential electricity demand. *Renew. Sustain. Energy Rev.* **2014**, *37*, 265–272. [CrossRef]

13. Witherden, M.; Rayudu, R.; Tyler, C.; Seah, W.K.G. Managing peak demand using direct load monitoring and control. In Proceedings of the 2013 Australasian Universities Power Engineering Conference (AUPEC 2013), Hobart, Australia, 29 September–3 October 2013.

14. Zoha, A.; Gluhak, A.; Imran, M.A.; Rajasegarar, S. Non-intrusive Load Monitoring approaches for disaggregated energy sensing: A survey. *Sensors* **2012**, *12*, 16838–16866. [CrossRef]

15. Hart, G.W. Nonintrusive Appliance Load Monitoring. *Proc. IEEE* **1992**, *80*, 1870–1891. [CrossRef]

16. Hosseini, S.S.; Agbossou, K.; Kelouwani, S.; Cardenas, A. Non-intrusive load monitoring through home energy management systems: A comprehensive review. *Renew. Sustain. Energy Rev.* **2017**, *79*, 1266–1274. [CrossRef]

17. Zeifman, M.; Roth, K. Nonintrusive appliance load monitoring: Review and outlook. *IEEE Trans. Consum. Electron.* **2011**, *57*, 76–84. [CrossRef]

18. Ruano, A.; Hernandez, A.; Ureña, J.; Ruano, M.; Garcia, J. NILM techniques for intelligent home energy management and ambient assisted living: A review. *Energies* **2019**, *12*, 2203. [CrossRef]

19. Anderson, K.D.; Berges, M.E.; Ocneanu, A.; Benitez, D.; Moura, J.M.F. Event detection for Non Intrusive load monitoring. In Proceedings of the IECON 2012-38th Annual Conference on IEEE Industrial Electronics Society, Montreal, QC, Canada, 25–28 October 2012; pp. 3312–3317.

20. Pereira, L.; Quintal, F.; Gonçalves, R.; Nunes, N.J. SustData: A public dataset for ICT4S electric energy research. In *ICT for Sustainability 2014(ICT4S 2014)*; Atlantis Press: Paris, France, 2014; pp. 359–368.

21. Yang, C.C.; Soh, C.S.; Yap, V.V. Comparative study of event detection methods for nonintrusive appliance load monitoring. *Energy Procedia* **2014**, *61*, 1840–1843. [CrossRef]

22. Weiss, M.; Helfenstein, A.; Mattern, F.; Staake, T. Leveraging smart meter data to recognize home appliances. In Proceedings of the 2012 IEEE International Conference on Pervasive Computing and Communications, Lugano, Switzerland, 19–23 March 2012; pp. 190–197.

23. Lin, Y.H. Design and implementation of an IoT-oriented energy management system based on non-intrusive and self-organizing neuro-fuzzy classification as an electrical energy audit in smart homes. *Appl. Sci.* **2018**, *8*, 2337. [CrossRef]

24. Makonin, S.; Popowich, F.; Bajic, I.V.; Gill, B.; Bartram, L. Exploiting HMM Sparsity to Perform Online Real-Time Nonintrusive Load Monitoring. *IEEE Trans. Smart Grid* **2016**, *7*, 2575–2585. [CrossRef]

25. Kong, W.; Dong, Z.Y.; Hill, D.J.; Ma, J.; Zhao, J.H.; Luo, F.J. A Hierarchical Hidden Markov Model Framework for Home Appliance Modeling. *IEEE Trans. Smart Grid* **2018**, *9*, 3079–3090. [CrossRef]

26. Wang, H.; Yang, W. An iterative load disaggregation approach based on appliance consumption pattern. *Appl. Sci.* **2018**, *8*, 542. [CrossRef]

27. Chang, H.H.; Lin, L.S.; Chen, N.; Lee, W.J. Particle-swarm-optimization-based nonintrusive demand monitoring and load identification in smart meters. *IEEE Trans. Ind. Appl.* **2013**, *49*, 2229–2236. [CrossRef]

28. Harell, A.; Makonin, S.; Bajic, I.V. Wavenilm: A Causal Neural Network for Power Disaggregation from the Complex Power Signal. In Proceedings of the ICASSP 2019—2019 IEEE International Conference on Acoustics, Speech and Signal Processing (ICASSP), Brighton, UK, 12–17 May 2019; Institute of Electrical and Electronics Engineers Inc.: Piscataway, NJ, USA, 2019; pp. 8335–8339.

29. Massidda, L.; Marrocu, M.; Manca, S. Non-intrusive load disaggregation by convolutional neural network and multilabel classification. *Appl. Sci.* **2020**, *10*, 1454. [CrossRef]

30. Wu, Q.; Wang, F. Concatenate convolutional neural networks for non-intrusive load monitoring across complex background. *Energies* **2019**, *12*, 1572. [CrossRef]

31. Georges Hebrail UCI Machine Learning Repository: Individual Household Electric Power Consumption Data Set. Available online: http://archive.ics.uci.edu/ml/datasets/Individual+household+electric+power+consumption?__hstc=262827539.79c1031e30e381d4e6e7812888505494.1474848000158.1474848000160.1474848000161.2&__hssc=262827539.1.1474848000161&__hsfp=1773666937 (accessed on 3 May 2020).

32. Kelly, J.; Knottenbelt, W. The UK-DALE dataset, domestic appliance-level electricity demand and whole-house demand from five UK homes. *Sci. Data* **2015**, *2*, 150007. [CrossRef]

33. Brownlee, J. How to Load and Explore Household Electricity Usage Data. Available online: https://machinelearningmastery.com/how-to-load-and-explore-household-electricity-usage-data/ (accessed on 15 February 2020).

34. Breiman, L. *Classification And Regression Trees*; Routledge: London, UK, 2017; ISBN 9781315139470.

35. El Mahrsi, M.K.; Vignes, S.; Hébrail, G.; Picardy, M.L. A data stream model for home device description. In Proceedings of the 2009 3rd International Conference on Research Challenges in Information Science, Fez, Morocco, 22–24 April 2009; pp. 395–402.

36. Álvarez Menéndez, L.; de Cos Juez, F.J.; Sánchez Lasheras, F.; Álvarez Riesgo, J.A. Artificial neural networks applied to cancer detection in a breast screening programme. *Math. Comput. Model.* **2010**, *52*, 983–991. [CrossRef]

37. García Nieto, P.J.; Alonso Fernández, J.R.; Sánchez Lasheras, F.; de Cos Juez, F.J.; Díaz Muñiz, C. A new improved study of cyanotoxins presence from experimental cyanobacteria concentrations in the Trasona reservoir (Northern Spain) using the MARS technique. *Sci. Total Environ.* **2012**, *430*, 88–92. [CrossRef] [PubMed]

38. Liu, Q.; Kamoto, K.M.; Liu, X.; Sun, M.; Linge, N. Low-Complexity Non-Intrusive Load Monitoring Using Unsupervised Learning and Generalized Appliance Models. *IEEE Trans. Consum. Electron.* **2019**, *65*, 28–37. [CrossRef]

39. Lin, Y.H.; Tsai, M.S. Non-intrusive load monitoring by novel neuro-fuzzy classification considering uncertainties. *IEEE Trans. Smart Grid* **2014**, *5*, 2376–2384. [CrossRef]

40. Kamat, S.P. Fuzzy logic based pattern recognition technique for non-intrusive load monitoring. In Proceedings of the 2004 IEEE Region 10 Conference TENCON, Chiang Mai, Thailand, 24–24 November 2004; Volume 100.

41. Bonfigli, R.; Felicetti, A.; Principi, E.; Fagiani, M.; Squartini, S.; Piazza, F. Denoising autoencoders for Non-Intrusive Load Monitoring: Improvements and comparative evaluation. *Energy Build.* **2018**, *158*, 1461–1474. [CrossRef]

42. Kim, J.; Le, T.T.H.; Kim, H. Nonintrusive Load Monitoring Based on Advanced Deep Learning and Novel Signature. *Comput. Intell. Neurosci.* **2017**, *2017*. [CrossRef] [PubMed]

43. Zadeh, L.A. Fuzzy sets. *Inf. Control* **1965**, *8*, 338–353. [CrossRef]

44. Cannon, R.L.; Dave, J.V.; Bezdek, J.C. Efficient Implementation of distinct the Fuzzy c-Means Clustering Algorinthms. *IEEE Trans. Pattern Anal. Mach. Intell.* **1986**, *PAMI-8*, 248–255. [CrossRef]

45. Dunn, J.C. Well-Separated Clusters and Optimal Fuzzy Partitions. *J. Cybern.* **1974**, *4*, 95–104. [CrossRef]

46. Bezdek, J.C. Objective Function Clustering. In *Advanced Applications in Pattern Recognition*; Springer: Boston, MA, USA, 1981; pp. 43–93.

47. Gustafson, D.E.; Kessel, W.C. Fuzzy Clustering with a Fuzzy Covariance Matrix. In Proceedings of the IEEE Conference on Decision and Control, San Diego, CA, USA, 10–12 January 1979; pp. 761–766.

48. Gath, I.; Geva, A.B. Unsupervised Optimal Fuzzy Clustering. *IEEE Trans. Pattern Anal. Mach. Intell.* **1989**, *11*, 773–780. [CrossRef]

49. Liu, H.C.; Yih, J.M.; Wu, D.B.; Liu, S.W. Fuzzy possibility c-mean clustering algorithms based on complete Mahalanobis distances. In Proceedings of the 2008 International Conference on Wavelet Analysis and Pattern Recognition, Hong Kong, China, 30–31 August 2008; Volume 1, pp. 50–55.

50. Killick, R.; Fearnhead, P.; Eckley, I.A. Optimal Detection of Changepoints With a Linear Computational Cost. *Source J. Am. Stat. Assoc.* **2012**, *107*, 1590–1598. [CrossRef]

51. Salerno, V.M.; Rabbeni, G. An extreme learning machine approach to effective energy disaggregation. *Electronics* **2018**, *7*, 235. [CrossRef]

52. Le, T.T.H.; Kim, J.; Kim, H. Classification performance using gated recurrent unit Recurrent Neural Network on energy disaggregation. In Proceedings of the International Conference on Machine Learning and Cybernetics, Jeju, Korea, 10–13 July 2016; Volume 1, pp. 105–110.

53. Alkhulaifi, A.; Aljohani, A.J. Investigation of deep learning-based techniques for load disaggregation, low-frequency approach. *Int. J. Adv. Comput. Sci. Appl.* **2020**, *11*, 701–706. [CrossRef]

54. Yan, K.; Li, W.; Ji, Z.; Qi, M.; Du, Y. A Hybrid LSTM Neural Network for Energy Consumption Forecasting of Individual Households. *IEEE Access* **2019**, *7*, 157633–157642. [CrossRef]

55. Fagiani, M.; Bonfigli, R.; Principi, E.; Squartini, S.; Mandolini, L. A non-intrusive load monitoring algorithm based on non-uniform sampling of power data and deep neural networks. *Energies* **2019**, *12*, 1371. [CrossRef]

56. Valenti, M.; Bonfigli, R.; Principi, E.; Squartini, S. Exploiting the Reactive Power in Deep Neural Models for Non-Intrusive Load Monitoring. In *Proceedings of the International Joint Conference on Neural Networks, Rio de Janeiro, Brazil, 10–13 July 2016*; Institute of Electrical and Electronics Engineers Inc.: Piscataway, NJ, USA, 2018; Volume 1.

MDPI
St. Alban-Anlage 66
4052 Basel
Switzerland
Tel. +41 61 683 77 34
Fax +41 61 302 89 18
www.mdpi.com

Energies Editorial Office
E-mail: energies@mdpi.com
www.mdpi.com/journal/energies